UPROOTING

UPROOTING

Marchelle Farrell

CANONGATE

First published in Great Britain in 2023
by Canongate Books Ltd, 14 High Street, Edinburgh EH1 1TE

canongate.co.uk

1

British Library Cataloguing-in-Publication Data
A catalogue record for this book is available on
request from the British Library

ISBN 978 1 83885 867 4

Typeset in Apollo MT Std by Palimpsest Book Production Ltd,
Falkirk, Stirlingshire

Printed and bound in Great Britain by Clays Ltd, Elcograf S.p.A.

Contents

Dormancy

Viburnum I   3
Primrose   21
Hellebore   43

Awakening

Blackcurrant   67
Judas Tree   91
Wisteria   114

Blossoming

Rose   137
Crocosmia   160
Verbena   180

Harvest

Mexican Hydrangea   205
Asters   227
Salvia   248

Epilogue

Viburnum II . . .   269

Acknowledgements   276

For our grandmothers

All flourishing is mutual.
– Robin Wall Kimmerer

Dormancy

Viburnum I

I WAKE AND do not know where I am. My eyes open on an unfamiliar scene of sheep grazing in a field, naked trees rising beyond them, stretching to the sky. The green folds of the land lie like crumpled velvet, a blanket softly draped around the seat of the house, trailing away up the valley. Birds call in the faint light: owls, crows. Last night I watched the full, cold December moon rise over the trees; now the sun chases her. In the stillness of this first morning I can hear the ever-changing song of running water. All is green, peaceful, idyllic. I seem to be gazing on this freshly wrought diorama of the English pastoral from our old bed. It takes a moment to remember, to piece together the meaning of this view through undressed windows framed by stacks of boxes: this is home.

I stretch and shuffle to sit up in bed, feeling the ache in my shoulders and lower back waking up alongside my awareness, my eyes transfixed on the scene through the windows opposite. Sky, field with the track to our neighbour's house running through it, hidden by the trees, our boundary hedge. I am awake but this feels like dreaming. A movement catches my eye, and I notice a brown figure picking its way across the far side of the field, tiny from this distance. When it raises its head to look around, I see that it is a deer. After a few moments of stillness it bounds away into the trees and I realise I have been holding my breath.

I slide out of bed and walk carefully through the sleeping house, bare feet on wood and stone, and step out onto the earth of the garden. The cold jolts sluggish bed-warm senses awake faster than any mug of coffee, the edges of myself sharpened and glistening in the dew. I am drawn to the stream, just here where a tiny waterfall cascades down a series of small pools before disappearing underground again. Listening to the water, I stare a while: at this stream, at this stone forming channels and terraces and a cottage, at this garden and the countryside beyond. I rest my hand on the local stone from which these terraces are carved, the same colour as my winter-sallow skin, and feel the tension of moving house die as I let these rocks carry my weight. I sit in this picturesque English country garden which is mine now. I try out the word on my tongue and it feels tentative and strange. Home.

What is home? I have been trying to make sense of the word, to find some belonging in it since I left my child-hood one in Trinidad and Tobago twenty years ago. The place I grew up in, that old colonial, tropical paradise, seemed to my unfurling teenage self unbearably narrow and small. There seemed no room beneath the tangled banyan roots, or on the burnt sugar-cane flats, or in the dripping, jewel-green valleys, for the wide horizon of ambition. On an island easily traversed within a day, there felt no scope to fully, imperfectly, grow into myself. I became claustrophobic; felt trapped by the confines of my vision. With its rainforest floor choked with the fallen branches of ancient family trees devastated by the blight of colonialism, for my generation there could be no risk of misdirected growth. Too much had already been lost, or stolen.

I was unsettled, a nameless something churning at my core, threatening the very ground beneath my feet. A ground that never felt solid, shuddering always with after-shocks that reverberated through my fractured bloodline, like the tremors we occasionally felt from the fault lines deep in the earth below. I left the islands that birthed me in search of secure grounding. Like wind-rushed genera-tions before me, I sought the Motherland.

At first I drifted, as ambivalent about England as it seemed about me, my inability to settle on this indifferent land not a concern. I took my free-floating as freedom. I had felt compelled to leave it, but my idea of home still lay in the emerald hills and turquoise seas that formed the backdrop to my childhood. A Caribbean oasis emerging like a mirage; hot air wavering over melting tarmac at the end of each parched transatlantic flight.

Then, slowly, things happened to ground my heart here in England too. Marriage, one child, another. Each loving kiss a new rhizome seeking to bed down in impervious English soil. As my family grew, so did a yearning to settle and take root, to be held, and grow strong in belonging. The idea of home began to leave me confused, my heart torn. I bandied the word about: I left home to return home, I missed home while being at home. But no soil beneath my feet felt like where I fully belonged, like where I could be entirely received in my fractured, imperfect parts.

Displaced only daughter of generations of dispersed people, I am the last seed adrift on a centuries-old breath. I uneasily spiral through time and place and fragmented memory. But I grow tired. In the garden of this place we have landed in, this place we hope to make our home, I rest at the edge of the stream, gazing at the dawn breaking over the swaying wood. The sky lightens through pinks and oranges into a pale winter blue. A group of birds takes flight from the trees, calling loudly, silhouetted against the morning sky. I listen to the chorus of the water and the wind and the birds and they sing to me of the possibility of belonging. It sounds like peace.

***

I felt peace come over me on our descent into the valley. The road narrowed to a winding tunnel between the trees, barely wide enough for two cars to pass. As my husband, Oli, and I drove slowly through the filtered, autumnal light I felt a spreading calm and a simultaneous thrill. Something within me was reverberating with the rightness of this place, and a part of me found that terrifying. As Oli frowned and concentrated on the sharp turns, I gazed up at the tree canopy, leaves interlaced from either side of the road. The steeply twisting journey through the green tunnel made me think of the drive across Trinidad's Northern Range to reach the North Coast beaches, a weekly Sunday pilgrimage every dry season of my childhood. The similarity was unexpected, and comforting. I half imagined we might glimpse the glittering sea as we rounded the last turn. Instead, the small collection of houses at the bottom of the hill shimmered as we caught sight of them, waves of honey stone resplendent in the light. We had arrived at the golden hour on a perfect October day, last viewing in a day packed with them. We were headed just outside Bath, to the kind of rural English setting in which I had never really dared imagine myself.

It was a small village, a few hundred residents maybe. A mix of housing, from council to manor, along one small high street. There was a village school, a busy pub, a community bus, and a shop run by volunteers as a co-operative. We passed a poster on the parish noticeboard for a night of Aretha Franklin's music next to one about rewilding. The village lay in a Conservation Area, in an Area of Outstanding Natural Beauty, and contained several Sites of Special Scientific Interest; it was as picture-perfect-rolling-hills-chocolate-box beautiful as all those titles suggest. Yet behind the perfectly preserved facade, there was a sense of a thriving community, something very vibrantly alive. People smiled and gave curious greetings as we parked and wandered through the village. They clearly noted us as outsiders, in the small-community way that I recognised from the neighbourhood of my own childhood, but were friendly nonetheless. I did not feel unwelcome. My senses were razor-sharp to this, for it was belonging that I was searching for, a community to call my own that I needed. I was going out on a limb here, a Black woman in the green of the White English countryside.

The first thing we saw was the garden. Heads full of all the other houses we had already viewed that day, we pulled into the drive, tyres crunching on gravel, and got out. Next to the driveway was a lawn, rimmed by a stream, bordered by a native hedge. A field lay beyond, a wood rose behind that, hugging the edge of the valley in which the house sat. The stream was shallow, and babbled merrily over its gravel base. At one point there was a bridge over it to a little patio framed by a pair of small trees and edged with what looked like black grass. On the other side of the lawn was a bed with a tree that Oli recognised as a quince, mounds of geraniums, and a neat box hedge that offered a glimpse of garden beds beyond. My heart was pounding. I was not sure I had ever seen anything so enchanting placed almost within my grasp.

I hardly remembered what I saw that day, my mind a jumble of fleeting images with an overall impression of overwhelming rightness. Yes. Yes. Yes. Yes, this was the place for us. There was old local stone, terrace after terrace carved out of the steeply sloping land that curved all around the house, the sound of water everywhere as the natural spring that emerged from under the roots of the ash tree at the top of the garden made its way down through the channels that had been created for it. I made my way through the interconnected garden terraces as through a series of rooms. I was besotted.

I barely noticed any detail about the actual house, bar that there were quirks. There would be issues and challenges aplenty in this haphazardly extended, centuries-old home, but the main feeling that I took with me of the building was one of embracing warmth. The jewel of the house was the garden room. A small orangery built in flawless local stone whose blue glass roof shone like a diamond in the estate agent's listing. I saw myself sitting there in winter, one more tropical being among the houseplants, drinking in the season's weak light. 'It is lovely,' I repeated, inadequately. I was astounded this lay within our budget. I was overwhelmed with hope. By the time we had gone to bed that night we had already put in our offer.

Monday morning arrived. Midway through my first meeting of the day the call came: our offer had been accepted. I looked back through the door at colleagues continuing the discussion I had already forgotten as the news that seemed almost too good to believe reverberated a death-knell on that life. I heard the sound of the stream welling in my head over the estate agent's excited voice, and as joy bubbled through disbelief I sensed that something profound had begun.

***

A sprig of pink flowers unexpectedly lands at my feet. It is nearly the winter solstice, and to my mind the dark midwinter means that all must be cold and dead, like my soul feels after the onslaught of weeks of dour wet. But here I am, a few days after uprooting from our old life and moving into our new house, still alive and wrestling bags of shopping in through the front door, every piece of the complex puzzle of relocating a family at short notice and with a tight deadline having eventually fallen into place as if by magic. The last few weeks have been such a string of serendipities and lucky synchronicities that, despite my misgivings about how I will fit in and adjust to life in the English countryside, despite my unnerving doubts that this could possibly be the right life for me, I cannot shake the feeling that we are somehow meant to be here. And here they are on the floor before me, delicate and small, but unmistakably fresh flowers. Their pale pink colour seems out of place in the grey and brown of an English winter. A faint flush betraying ongoing life on the seemingly dead cheek of the garden.

I put down my bags and pick up the flowers to look more closely. As I draw them to my face I am hit by their honeyed scent. The smell is warm and sweet, and tugs at the edge of my memory. But the present intrudes, children storming noisily past, reminders shouted to take off muddy wellies. I wade through the sea of discarded coats and boots, and glance out the door to see where the flowers might have come from. For the first time I notice a shrub tucked into a corner of the building next to the front door. It is midwinter, and yet its branches are covered in tiny bunches of these flowers. For all the years I have lived in the UK, I do not believe I have ever seen such a thing. Or I have never noticed. But here they are, tiny bunches of flowers held on the ends of dead-looking stems. The garden has offered us a welcome bouquet.

The garden called us here, but in the tumult of our arrival we have not been in the garden. Everything is cold and muddy when I venture out later, lured by the possibility of more flowers. I wander in my coat with a coffee, trying to orientate myself in the space, tracing and retracing its contours beneath my feet.

At the top of the garden is some patchy grass, to one side of which stands a fully grown ash. Its bare silhouette of curving branches is a magnificent crown to the space. Down a short flight of steps the stream emerges, welling from a small spring almost hidden by creeping ground cover beneath the bare branches of a small acer. The water bends under the path, and then back again in an s-shape that meets the path's curves for infinity. The path changes underfoot from shingle, to stone, to brick. They both run through what seem to be the main ornamental terraced beds of the garden, the outlines of some naked shrubs on the left partially screening three pairs of veg beds made of old scaffold boards, which steeply rise to the back boundary native hedge.

Another flight of sharply winding stone steps leads down to the main patio, a rectangular space paved and walled in stone, and nearly level with the conservatory. Beside the steps, the stream tumbles down a series of moss-lined miniature waterfalls through a cascading set of small pools, then runs in a rill along one side of the patio before disappearing underground. As the stream crosses the space, it is bridged by sleeper and gravel steps edged by small square beds, some of which seem to contain herbs such as sage and bay. These wide steps climb away from the paving with sloping, winter-scruffy beds on either side. In the season's emptiness, this bit of the garden above the house seems mostly made of water and stone, and the flowing babble constantly resounds through it all. Contrasting with my hot, bitter coffee, the air is cold, and almost sweet. I swallow deep, gulping breaths of it.

Stepping down from the patio, the path changes to the crunching gravel which surrounds the house on all sides. The front of the building, facing onto the main village road, is clad in the branches of a climber I do not yet recognise, and shielded from the road by the brown beech hedge which forms the other boundary of this long, thin, steep space. Behind the house, flanked by a series of neat box balls, the gravel path leads to a circular stone seating area, which runs onto a narrow, rectangular lawn that spans the length of the house. Here there is a Wendy house, and a rabbit hutch and run, features that most excite the children. At the far end of the mossy lawn is a straight box hedge, next to a smaller circular patio that mirrors the first. This sits at the top of semi-circular steps which lead down under the quince tree to the larger lawn by the driveway at the very bottom of the garden, the first thing we saw on our arrival here.

I slowly circle the garden, still tired from the effort of moving house, thoughts drifting as I seek signs of the beauty that I remember from our visits not that long ago. Eventually I am drawn to the main ornamental terrace just above the house. Here I stare at the empty beds filled with grey clay, willing the soil to share its secrets with me. This bit of the garden looks like it has been worked at the most, creating the beautiful curves of stream and path and terrace that lure me to the space. And yet the earth looks barren, the beds perhaps cleared in an over-zealous late autumn tidy? I know that it is winter and it should not surprise me, but it feels so at odds with my memory of a lush and fertile space from our visits only a few weeks ago that it is jarring. In my mind's eye the garden was beautiful, with dense vegetation and heaps of flowers; but this looks distressingly like a wasteland.

Despite having lived through twenty English winters now, it still takes me by surprise, the way that all the lush growth of summer just ceases to exist every winter. Logically, I know that life continues underground, above ground even in the evergreen plants, like the pine tree the children and I made it our first concern to find and put up in the conservatory, the ivy growing in the hedges that I have not yet got round to weaving into a wreath for our new front door. But somehow winter has always felt like dying.

The winter garden scene before me is uncomfortable for another reason: the desolation that I hoped to leave behind now reflected in this ugly mirror. I do not want to remember the infertile feelings of the life that I have fled, but here they are all around me. The cold, empty scene haunts me. I wonder how much can be left behind when you leave a place, and how much you carry within you.

I have carried much within me. Five years of longing in a scarred and empty womb. In sterile compensation, I had invested all my creativity as a doctor, a therapist, a hard-working healer and clinician. I had poured myself into this work in the hope it would hold me, keep my broken parts whole even as I healed others. But the vessel was cracked, and could not hold all that I needed it to. Instead I had arrived, depleted, at the pinnacle of my career. I had been finally appointed consultant, only to find the promised uplands of seniority a burnt-out landscape, its resources drained. And my attempts at making a home for myself there thwarted by the inescapable barrier of my race.

When, after the long drought of infertile years, I finally had my children, caring for them had birthed a latent desire in me. During my maternity leaves, I had felt pulled by their needs to explore a different way of thinking about our life. But I could not sustain it in the eviscerating pace of our old, dual-medical-career life, and it had been near-obliterated by the unforgiving return to work.

As I hunch my shoulders to keep off the cold, I begin to realise that I was seeking something that I thought I had felt in the beauty and abundance of this garden when we viewed it in autumn. Something that felt wrapped up in the idea of home. But standing in the winter gloom, I can see only the ugly feelings and memories I had unthinkingly packed and brought with me. My musings heavy and glum as the soil beside the path, I see myself in the garden, and I do not like what it shows me.

I stare at this fallow land, and it silently stares at me in turn. That career that I drained myself to grow seems dead now. I have given all that up to move here, in a leap of faith, a grasp at creating something new. It was a willing escape, but the bleak stare of the garden makes me wonder what patterns I have been unconsciously repeating, how freely chosen this uprooting was in the end.

I take a deep breath. I am not dead. I am in a state of suspended animation; this is what dormancy looks like. I am fully spent from the intense years of longing and mothering and work. I want to hibernate, bury myself in a pile of leaves and let the endless rending responsibilities and obligations slowly fall away until some future restored state of spring. I am desperate to lie inert in the soil. And yet, giving up even briefly the strong sense of identity and purpose that has kept me going for so many years is terrifying. Through all the upheaval and turmoil of training, my medical identity was my home. It has hurt me deeply, lashed scars across my inner landscape in the relentless drive to keep going, chained me to a career treadmill that I have longed to escape, but who am I without it? Who am I if I am free? I am not sure I know.

I walk round to the front of the house, and look for the flowering shrub that so surprised me. It is still there, not a figment of my disoriented imagination, still holding its tiny, pale pink bouquets at the ends of bare brown branches. I stare at it, wondering whether other things flower here at this time of year too. How have I never noticed that there could be this kind of vibrant life in the dead depths of winter? I have read the advice that I should get outdoors as much as possible in the winter, to combat the descent that the season brings. Yet I have lived in buildings and cities, behind walls and heavy curtains, under artificial lighting with the heating turned up high. Winter has been hard for me to love, and I have not felt at home in it. Perhaps I have not wanted to.

A slight shift in the air wafts the sweet scent of the blooms towards me; again I am taken aback; they smell larger than they look. I wonder what other signs of life might surprise me in this bleak season.

The stream seems to chatter more loudly, and interrupts my thoughts. I walk over to sit on the stone bench beside it, and watch the water for a while. It is lined with moss and tiny plants with delicate, lacy leaves, green platforms forming three miniature waterfalls into a small pond. It reminds me of the deep pools with tumbling falls down fern- and moss-lined rock that I have hiked to and bathed under in the tropical forests of my childhood, but on a scale so small it must be for fairies. Crystal clear as the Argyle Falls, three pools separated by cascading water, ropes to guide the climb from the bottom pool to the top, rope swings to carry a laughing, tumbling body out into the water below. Water green with the reflection of the forest that meets overhead, gold with the fragments of sunlight that make it through the canopy to glint on the pools below. This water in my stream is also moss green and winter-sun spangled. But so much colder – I cannot hold my hand in it for long, and laugh and shiver at the thought of full immersion. As I sit, the two landscapes superimposed in the double exposure of memory, a robin alights on one of the stones. I hold my breath and stare at him drinking from and bathing in the stream. He looks at me curiously, and hops closer, unafraid. I think he is trying to figure me out, work out what kind of human I will be in this space. I wish he could tell me.

***

The winter solstice comes, a day that has become my favourite in the year. I love the paradox that it holds. It is the shortest day, the darkest time of the year. Yet even in the heart of that bleakness, it holds the memory of summer. After today, no matter how desolate the rest of winter, the light will build again. It serves as a reminder that this season of cold, darkness and death shall pass, and warmth, life and light return.

For the last few years we have tried to mark the winter solstice. It heralds the start of a season of celebration in our family, the first spark that ignites a fire of joy that blazes into the new year. Solstice, Christmas, New Year's Day and my father-in-law's birthday, our wedding anniversary, and finally my son's birthday – it is a busy, wonderful few weeks. Today I ready the logs and kindling for a warm fire in the grate, find the candles, and arrange them in the conservatory. As dusk falls, candles glow in a makeshift centrepiece for this darkest day that marks the return of the light. It all feels poignant, in this room with a glass ceiling that lets the waning moon stream in, as I hope that this move marks our shift to a season of light.

***

Christmas morning arrives, and I am back in the garden, coat over my pyjamas, muddy boots on bare feet. The boots were box-fresh just a few weeks ago, bought in preparation for our new country life, and with every patch of mud that I encounter, I am grateful to my unusually practical past self. Despite being in our new house for only a fortnight, we are hosting this year's festivities. This is bittersweet; it is the first time that we have had the room to consider hosting, and also the first time my in-laws, who now live nearby, are not up to the task. My father-in-law was diagnosed with an often incurable cancer the week that we viewed this house. He had surgery to remove the malignant growth on the day that we moved in. He is recovering well, but there has been an unspoken seismic shift in the family dynamic. Aftershocks rumble still, but the landscape has changed; my husband, the eldest son, will be the one carving at this Christmas table.

In my pile of presents, laid in a mess of wrapping under the tree in the conservatory, there are garden books and tools, and packets of seeds with beautiful, unfamiliar names. My favourite is cosmos, as it makes me imagine I could grow the heavens above in my garden below. I am out here looking for something with which I can decorate the dining table, and finding it as bare and ugly as the last time I looked at it, I vow to start sowing my new seeds as soon as I can. We have a small poinsettia inside, given to us as a present for our new home, with which we can make do. Battered by the midwinter move, it is a small, withered thing, a pale imitation of the flaming bush that stood in my grandmother's front yard every year at this time. She glowed proudly on the porch, her house decked out for the festivities, scrubbed and shined from top to bottom, new curtains screening the sliding glass doors into the house behind her, tree lights twinkling through the gauzy drapes.

I look at our tree – the biggest I could fit into our car with the kids, the branches tickling our faces as it stretched up between the seats, now loaded with all our baubles, some bought new specially for our first Christmas in our new home – and feel pleased that though the rest of the house is not dressed to her standards, she would have approved of the effort I put in there. My eyes slide to the sad poinsettia on the dining table next to it. The vibrant memories of my past – steam rising from cornmeal pastelles unwrapped from banana leaves; the sweetly tart taste of sorrel, a cordial made of hibiscus flowers; being made to put aside a plate of pelau in order to dance with an aunt to the lively music of a parang band – cast into pale relief the English garden centre's attempts to replicate tropical traditions.

I have never bothered to buy one of the ubiquitous potted poinsettias myself, shipped halfway across the world to stand in rows in supermarkets and garden centres, as certain a marker of the upcoming festivities as the plastic Santas and glitter-shedding angels. Whenever I have noticed them on hospital wards or coffee tables, they always seem to hate the confines of a winter house. And even if they manage to limp on through and survive the season, creating the particular routine of light and dark exposure that triggers the colour change so reminiscent of this time of year in the tropics is beyond me. I need to find a way to create my own festivities. I cannot simply fabricate my childhood home, I must grow something new. I look around this unfamiliar space and feel daunted at the prospect.

Eventually I find some poinsettia-bright twigs of dogwood in the hedge, and catkins from a shrub planted in a prominent spot next to the front gate. This shrub worries me; I have noticed others like it in neighbours' gardens and it should be in full evergreen flush. But this one is largely bare; what leaves it has left are dull, spotted and sickly. It feels half dead, which does not seem an auspicious response to our arrival in this space. The other day, I had pointed out this sorry-looking shrub to Oli; he dismissed my concern as superstition, and the mere effects of midwinter. Left alone with my morbid fears, I fret.

To curb the somber flow of my thoughts, I wander over to see our noisy stream. The constant flow of water welling from deep underground is the most lively thing in the dead winter garden, and is rapidly becoming my favourite thing about the space. Every day it chatters and babbles so much that I fancifully allow myself to half believe that a naiad of some kind lives here. As I walk towards it, I see a flash of bright orange dart towards a half-upturned pot at the bottom of the small pond. Something does live there; I am startled to realise that we have inherited a surprise goldfish. I sit at the edge of the pond awhile, wondering if the fish will be as curious about my presence as I am about theirs. It refuses to come out. I turn away to look back at the lights sparkling in the conservatory. I watch the children playing with new toys by the twinkling tree, Oli clearing the mess of one meal to make room for the next.

As I return indoors I brush past the shrub bearing the miniature bouquets of pink flowers that had surprised me on our arrival: *Viburnum x bodnantense 'Dawn'*, I have since learned. It probably needs pruning, but as it releases its scent with that familiar warm sweetness, I am fully transported. Another home, another steeply terraced garden, poinsettias aflame and a night-blooming jasmine in flower, honeyed scent thick against the goosebumps raised on my skin by the mild December chill of the tropical air.

I stand halfway through the front door, breath stolen by the strength of the memory of my childhood Trinidadian home. A shiver runs through me at the resonance, past chill raising present. That old home seemed dead and gone, yet here it is in all its warm familiarity. A home once loved, left and lost, revived in bright, shining memory in my strange new garden, where beauty blooms still in the cold darkness.

Primrose

MY SON STARTS school on his fifth birthday. An early January birthday seems a challenge at the best of times, but this year it feels particularly cruel. We are almost halfway through the school year and he is a newcomer entering a group of already-formed friendships, and it is his birthday. I remember the loving, diverse school we left behind and hope that this change will be worth it, that he will not come to hate me for this uprooting in years to come.

Standing alone watching the lively groups of other parents, I am nervous at the school gate. For all my smiling exterior, I feel insecure and reserved, the task of meeting people and forming new friendships filling me with dread. But that is exactly what I am here to do, so I try to let down the guard formed and strengthened over a lifetime, to keep myself soft and open, and return every smile. In the afternoon my first-day efforts are rewarded. I sent my son into class armed with cake, in hope of sweetening some of the difficulty of the day for him. The sight of the children emerging from the classroom waving cupcakes and trailing crumbs disarms us all effectively, a wave of relief as my son is kindly greeted with a chorus of happy birthdays and hopes that we will enjoy settling in here. The warm smiles around him make me hopeful that in time he might come to feel that he belongs here.

After the first viewing of what was to become our future home, I had returned alone to see the school. I parked in the middle of the village and walked through the pictur-esque autumn leaves, apprehension building with every crunch beneath my feet. The idea of living so close to my children's school filled me with joy. Not all of the school's students came from within the village – it was too small to populate a viable primary school on its own. Nevertheless, this pretty, rural outpost, the type of place that often seems a ghost town of second homes, was full of families with young children. My head swam with visions of the chil-dren playing wholesomely outdoors, and running between friends' homes in a few years' time; I wanted the freedom that I had felt in my own childhood for them. But I was worried about how friendly the experience would be. Rural Somerset was not renowned for its diversity; rural Britain held the stereotype of being hostile to outsiders. I had been made to feel vulnerable and unwanted because of my race in this country. I did not want my children's faces to be the only brown ones at their school; I remembered too well how cruel about difference children can be.

While walking along the wall of the school's perimeter, a sudden blast of familiar rhythm reverberated from the stone. I paused, doubtful of what I was hearing, too star-tled at first to recognise an old friend in this totally unexpected place. But here was calypso, melody swirling and swaying above this golden-green, quintessentially English scene, from what turned out to be a rehearsal for the school play. As my anxious heart rate slowed to match the reassuring rhythm of my childhood, I beamed with delight when shown the spectacle of a hundred-odd children dancing to this music I had known from the womb. I did not ask the question directly, but I did not need to – I felt the clear wash of relief as I gazed over a scene more diverse than I had expected, including a few children who seemed mixed-race like my own. How this relatively disparate diaspora had all landed here in a tiny village school in the West Country a mystery; some ancestral spell for belonging come to fruition.

***

We walk through the garden as we leave and return, on the way to and from school. Climbing the terrace steps to the small gate that puts us out onto the road at the top gives me at least four chances every day to notice how dispiriting the January garden is. The brief spark of the possibility of joy in winter lit by the flowering viburnum that greeted us on our arrival had soon faded, fizzled in the soggy remnants of brown petals left behind. I always find the dissipation of light and warmth after Christmas challenging, and here in the absence of city lights the contrast seems magnified. Everything is wet, and muddy, and cold, and dark, and I begin to realise that the blessing of living in the countryside, so close to the seasons, might be a curse in these most dismal days of the year.

I begin to develop an irrational hatred for the garden's waterlogged soil, speckled with stones and rubble, feeling almost as if it is being purposefully loathsome in its winter nakedness. I glare at the beds as I cross the garden, willing them to remind me of the beauty that so charmed us in the autumn. They sit silent, unyielding, even the evergreen shrubs that frame either end of this main ornamental terrace which we cross many times a day appearing a sullen washed-out green under the leaden winter light. There are a trio of shrubs that sit above the stream's cascading pools, clipped into rough mounds which screen the veg beds from sight when viewed from the house. I brush against the foliage of the nearest whenever I climb the terrace steps, and vow to dig the ugly beast out come the spring.

New neighbours tell me that this garden was a labour of love for the previous owner, how hard she worked to carve it out of the hillside. The gifts of that labour and love are evident in the rock walls and the carefully meandering stream, but the neglected beds continue to distress me, looking as if care has never reached the soil and reminding me of things I would rather leave behind and forget.

At first I avoid the space, moving as quickly as I can through it on the school runs and trips to the village shop, eyes averted. But while crossing the terrace one dreary morning, the garden's robin darts across the path ahead of me. Startled, I pause, and look to find it. The bird sits with its head cocked in the middle of the most dismal bed on the terrace. Sighing, I stop, and let myself really look at the garden. The bare twigs of a few deciduous shrubs stand stark against the earth, dotted with sparse lumps of the withered growth of likely perennials. The stream wanders noisily through its stone channels, the moss lining its course the only vivid life in the space. Without the plump green flesh of summer growth, the garden is all bones.

I crouch at the edge of the path, and picking up a handful of the unlovely clay of the garden beds I squeeze it into an impression of my fist. Retaining my hand's shape as I let go, it does not impress me with the sense of land that has been lovingly tended. Rolling the greige dirt more gently into fat sausages between my fingers, I think of the hands that impressed themselves on this land before mine. In this austere light, I can almost see the shadows of those who passed through this space before me, their echoes reverberating off the stones. This place was much loved once. What caused the love to leach out of the soil here? What made previous stewards of this place leave?

***

Rootless wanderer that I am, I should know. Tracing back through the generations before me is a tortuous trek across the seas, no one settled in place for much more than a generation before some displacement happened again. 'Go back where you come from', the common insult hurled at those who look like me. I spent my childhood wondering where exactly I would go.

I was only just born in the same country as my parents, my father having followed his siblings, nearly all eight of whom flocked from Trinidad to the bright lights of New York City like neotropical migratory birds. They flew like so many others to the land of milk and honey, lured by dreams of a better life, sold on the story that we all were, that the island was a third-rate, third-world, corrupt place where dreams withered in the midday heat. A decade in New York, he coaxed my mother to join him for a time. Until the last, heavily-pregnant moment, when she found herself returning to her childhood home for my birth, inexorably summoned by some unknown instinct to familiar ground.

Further back the upheavals are disorienting, their traces faint. My maternal grandmother's mother born on one Caribbean island, her father on another, some of their parents on different ones still, their parents coming from England, Scotland, parts of Africa. Which parts remain a mystery, documentation disappeared or destroyed. All that remains are echoes of stories of some enslaved and some free Black people being brought in various ways to this tropical archipelago. The branches of my maternal grandfather's family tree are even more obscured by the mists of time. As the son of an unmarried woman and a married man, more is rumour and hearsay than fact, but veiled stories of Portugal and China live in his name, look out from his serious face in the photos left behind.

It is a similarly shrouded picture on the paternal branch of my family tree. My father's mother the daughter of those enslaved, stripped of themselves, then freed, moving from island to island in search of a self and a home. My paternal grandfather claiming to be the grandson of an Irishman, the source of our family name. A story dismissed by my dark-skinned father as a simplistic desire for connection with Whiteness, a denial of the painful reality of plantation slave name attribution. That was until he met his pale, green-eyed cousins.

Our history is branded on our skin, rusted manacles absorbed into our bones, with few living stories to ground us. Dead ancestors lie scattered beneath the sea. We float like seaweed.

I know some of the bones of the tales of how my ancestors were forcibly displaced, torn from their homelands and shipped across the sea to the Caribbean in acts of great cruelty, at high moral expense for all involved. Breathing, bleeding machinery at the frontline of Imperial agricultural practice. Names, languages, souls lost. Shackled attempts at rebuilding new selves from the ground up with the cane and cocoa on those jewel-green islands in a turquoise sea. Tiny islands, freighted with the heavy burden imposed by the gravitational density of Empire.

I know so little about the flesh of their lives. Entire people inexorably drawn by these migrations. Thrown together by the violence of Imperial upheaval to make me. Where were their homelands? What were they leaving behind? What welcome did they find? With only shadows to guide me, I struggle to understand what came to connect them all. Ghostly echoes remain, the question of home unanswered, living on in the unanswerable question of me.

This lack of rooted presence in a place haunts me. There were whispered stories of an Indigenous ancestor when I was a child, contemptuously slapped down by my beloved Anglophile grandmother, but we all saw the Carib features that she strenuously denied. The same ancient face that looked out from the dusty black and white photos of the original people of these islands in the local museum. For my own desperate need for attachment to place I clung to them, though I could not see them in my own reflection. My flesh may have been carved of many parts, but I was wholly Trini at the bone, and these Caribbean islands were where I belonged, as I believed my great-great-great-grandmother had before me.

And then I left.

***

As the seemingly endless weeks of January go on, I begin to see the emerging promise of spring in other gardens in the village. Everywhere, lips of green begin to break the muddy surface, speaking quietly but insistently of joy to come. In my garden, tucked away near the very bottom of the valley, in the frost-pocketed shade of the hills and trees around, where any weak winter sun touches only the garden's highest edges, I see few such signs yet. But then I notice the primroses.

Primula vulgaris, the internet tells me, is indigenous here. It belongs to this place. It is named for what I can see: it is the very first thing to flower in this new year. I read that the primrose represents eternal love, and is said to protect the home from evil faeries in Irish folklore. This gives me warmer feelings to the generous clumps of small, pale yellow flowers that begin to appear around the garden. Usually too insignificant a plant for my more exuberant taste, my heart lifts to see their bright yellow and green against the unrelieved winter gloom. My son absolutely loves them. He declares them his favourite flower, and on one relatively sunny weekend day when we are all tempted to linger outdoors, he carefully digs up a couple of small seedlings, which he places together with some other unidentified weeds in a pot to form his own garden. He nurtures them lovingly, picks the tiny flowers and offers me miniature bouquets. I sense that he understands something of these quiet flowers that I, foreigner, do not.

Holding the flowers he has given me, I am reminded of my grandmother, of the old fashioned 'granny' plants she loved and used to grow in pots and baskets on the verandah. Petunias and periwinkles, something of the similarly simple flower shapes transporting me for a moment to that earliest garden. I wonder if she would have loved the primroses, or deemed them too pale and uninteresting. But they are native here, and she loved everything that was native here, and denied her own belonging. I look at my English garden and think of her Trinidadian one and wonder if, in following the compulsion to leave there and settle here, I have done the same.

***

As a child, I first lived in a small annexe to my grandmother's home while my parents – still at university when surprised by my impending arrival – studied and worked to achieve financial independence. My mother built the annexe in my grandmother's garden when she became pregnant with me, an extra room to house her growing family. The garden that our room ate into was a standard plot for a detached, suburban Trinidadian bungalow of the time. The front garden was closely clipped grass with a generous ixora hedge spilling out over the fence. A few specimen shrubs dotted the small lawn: most memorable to my childish mind were the poinsettia, which blazed each Christmas, and my grandmother's attempt at growing English roses. They were sickly in the tropical heat, always with ant nests in their roots.

The back yard contained the kennel where the dog lived, and space for washing to be hung, or my paddling pool to be put out on the hottest days. Otherwise, it was mostly given over to trees: mango and avocado, orange and lime filling the modest space. I loved climbing them, trying to claim their fruit before the birds could. Our mango tree was not the best on the block, though. That honour was held by our neighbour's prolific Julie mango tree, whose heavily weighted branches would drip with sweetly abundant fruit over the fence into our yard every mango season. There would be such a glut of the ripe fruit that even I, who loved Julie mangoes more than anything, would be unable to keep up, and would start to find the sweet smell of those which fell to the ground and rotted under the tree overwhelming.

When I was eight, we moved briefly into a rented townhouse, then into an upstairs flat in the lush Port of Spain suburban valley of St Ann's and Cascade, a neighbourhood that I was to grow up thinking of as my home ground. The flat was in a compound of two or three adjacent houses divided up and converted into rentals in this popular area. The shared garden was mainly made up of lawn and tall, broad-leaved trees, maintained by an improbably fit, eighty-year-old gardener. His favourite trick was bringing me lengths of sugar cane from his own land and stripping it with his bare gums, toothless face alight with a huge grin at my amazement every time.

A few years later, and we were finally in our own garden, my parents having proudly managed to buy our family a home in one of the city's most beautiful suburbs. They had achieved the holy grail of the model we were all taught to aspire to by our colonial betters, where every man's home was his castle: they had arrived at home ownership. In the concrete urban sprawl of Port of Spain, the St Ann's and Cascade Valley gleamed a green oasis. It was home to the President's House, former abode of the British governor generals who claimed the island in colonial times, and tried to establish profitable plantations and keep the diverse and unruly population in check.

Many of the buildings erected in the British colonial era, wooden constructions built slightly raised off the ground, with wraparound verandahs and decorative fretwork in white and pastels so that they looked like elaborately iced cakes, still stood in the neighbourhood where my parents had bought our home. These homes were a decidedly Trinidadian mix of all the complex architectural heritage that the island harboured. I thought they were the most beautiful.

The house my parents bought had echoes of that architecture, standing on stilts on its uneven site, with an upstairs verandah, but it was a modern building of concrete breeze blocks. My parents painted it green and white, installed a cobalt kitchen, and tiled the floors in terracotta interspersed with tiles handmade by a local potter who had settled here from England after falling in love with the island. And then we moved at last into our forever family home, and set to creating a garden.

After a lifetime of living in others' spaces, no matter

how lovingly housed, it was a revelation to have a home of our own. A room to decorate according to my teenage tastes and pocket money. My bedroom, screened and cool in the heat of the day by a bank of giant ginger lilies growing to meet the roof just outside, was a space to contain teenage angst, and a regular roster of slumber parties. Our evolving garden one that could keep pace with the ambitiously expanding scope of my dreams. Held by the gentle folds of the valley around, I changed from a child into the burgeoning adult version of myself. I left at age nineteen to go to university in England, and though I imagined myself a grown-up who had flown free then, it was still my home. We had finally acquired the secure base from which I could bravely leap forth into the world.

And then, a few years later while I was overseas, my parents sold it. The reassurance to my anguished reception of the news was that it had been done for me, as the property market was in a wildly expanding bubble, and they had sold it for many times more than they paid for it. My logical, adult mind knew that this was an incredible move for my parents, especially my father, born as he had been in one of the island's historic but notorious neighbourhoods. Morvant Laventille, covering the hills rising to the east of Port of Spain, was where many of the enslaved made their homes when slavery was abolished, leading to the area once being known as Free Town. The birthplace of the steelpan, and a breeding ground for so much of the creative soul of the island. And in more recent times, stigmatised as the home of violent crime and profound poverty. He and his eight siblings had made their escape; he spoke very little about his childhood. I could only sense its echoes in his sheer determination to provide a different life for me.

The selling of our home was a coming together of decades of hard work with a moment of financially savvy serendipity that would ease all our futures. My parents' future was now economically secure. And they reassured me that, as their only child, so was mine. This was why they had done it: to create an inheritance. To build material generational wealth. Now they could afford to supplement the meagre living allowance of the academic scholarship that had allowed me to come to England; they wanted me to enjoy my time at university. They did not have to scrimp and save so much any more, and we could all travel to Europe on holiday now; it would be wonderful!

But the wailing child within me heard only that my devastation was my own fault. I was homeless in the world now, and in my neediness I had made it so. The beautiful forever home with the lush garden on the green hillsides of the Cascade Valley was gone, consumed by the greed of my ambitions.

***

The island of Trinidad was first settled by the Indigenous people of the Americas. The oldest archeological site in the Caribbean, with the most ancient human remains yet found in the region, exists in Trinidad, dating back nearly seven thousand years. As the closest of the Caribbean islands to the South American mainland, only about seven miles off Venezuela, it has acted as a transit hub for trade and exchange, and the migrations of people, for millennia.

The Spanish believed themselves to have discovered the island during Christopher Columbus's third voyage in 1498. They claimed it as their own, and renamed it La Isla de la Trinidad, for the trinity of hills that Columbus saw on his approach. The original people had called the island Iëre, Land of the Hummingbirds, after the large numbers of hummingbirds, which were viewed as sacred, who made their home there.

At the time of the arrival of the Spanish, the island was densely populated with a diversity of Indigenous tribes: Nepuyo, Carinepogoto, Warao. Names now almost lost to local vernacular. These tribes came from two different language-speaking groups, the Arawakan and the Cariban, resulting in their people eventually coming to be crudely known as the Arawaks and Caribs. Within fifty years of Spanish arrival, this richly diverse original population had been reduced from an estimated forty thousand people to a mere four thousand. People had been travelling to this thriving region for thousands of years, and yet in their relentless greed the arrival of Europeans brought genocide.

The Spanish came in search of gold and, even more valuable, souls. A string of governors was appointed, and repeated attempts at settlement made, but for a time they were driven off by strong resistance. A permanent foothold on Trinidad was eventually established, but the number of Spanish settlers remained small, and outnumbered by the locals, who continued to resist Spanish domination.

In order to tighten their grasp on the island, which occupied a strategic position for trade close to the Spanish Main, attempts were made to increase the island's population.

Roman Catholic planters from other islands were invited to settle in Trinidad by the Cedula of Population in 1783. This significant arrangement granted land to each Roman Catholic who would swear fealty to the Spanish Crown, and for each enslaved person that they brought with them. Unusually, it also granted land to *gens de couleur libre* or 'free people of colour', along with their slaves. This brought about a defining change in the population of Trinidad, and introduced a unique middle class of free, slave-owning Black people to the island's already complex ethnic mix. The mainly French planters, fleeing the turmoil of the French Revolution resounding through their colonies, brought with them their thousands of enslaved people and plantation culture, and began the mass cultivation of sugar, coffee, cocoa and cotton on the island.

Despite these attempts, the Spanish could not hold the island against other empires' greed, and eventually capitulated to the British in 1797, resulting in a colony that was British-ruled, with Spanish laws, dominated by French culture, and speaking a variety of languages. The French, Spanish and Indigenous languages combined to form the patois that was spoken alongside the introduced English until the latter's eventual dominance. Patois is now an endangered language, mainly sung in local Christmas carols, and spoken fluently by very few. British rule brought another wave of migration, with British and other European families coming to the island, as well as other populations of Black people.

The British governors tried to establish order in this 'Latin Mess' of a colony, but found the heterogenous population frustratingly unruly subjects. This led to the introduction of an extremely harsh Slave Code on Trinidad by its first British full-time governor, Thomas Picton. He would later achieve fame for his performance in the Napoleonic wars, but he first came to the notice of the British public for his cruel treatment of the enslaved population in Trinidad – an embarrassment for the British government at a time when the abolitionist movement was gaining traction. These Slave Codes introduced by the British throughout the Caribbean formally established for the first time different races of men, and instituted their hierarchy of worth. They justified the cruel treatment of Black and Indigenous people, who were deemed subhuman, and incapable of living independently.

The abolitionist movement brought further turmoil to Trinidad, and the island was used as an experimental 'model slave colony' to see how lenient the British could be about slavery without needing to commit to the expensive and inconvenient business of abolishing it altogether. The eventual abolition of slavery – formally in 1833, but entirely in 1838 after the controversial and much protested 'apprenticeship' system was ended early – brought further migrants. Venezuelan farmers came to manage the cocoa estates, as Trinidadian cocoa was at the time some of the most valuable in the world, and by the 1920s made up a fifth of global cocoa production. Agricultural workers came from China, and at the same time a programme was introduced to bring indentured servants from colonial India to manage plantations which the ex-enslaved people had understandably abandoned.

And so this tiny island in the Caribbean, with its mysterious geographical density that draws others from all corners of the world to love and covet it, became an enforced home to a diverse population once again, this time an uneasy melting pot of cultures crushed together by the black hole of Imperial gravity, and constantly on the brink of simmering over with the unresolved violence of its creation.

***

I am hoping the garden will teach me how to grow into belonging here, but this January is proving a hard time for learning its lessons. The bite of winter is sharp and deep. One day I find a photo of myself on my phone, taken by one of the children. It shows me standing on the frosty steps above the circular patio behind the kitchen, looking out at the lawn below. This is how I pass any of the time that I do manage to spend in the garden in these early weeks: walking around various bits of the space, looking at it.

The looking has various uses. I read that noticing where light falls is important in helping to understand which plants suit different locations. That seeing what is already there by way of weeds will tell me much about the condition of the soil. The few seedlings that I do see overwintering in the beds I eventually identify as fringed willowherb. This tells me that our soil is wet, and disturbed. I have a sense of knowing this already – the disturbance resonates with my own, and makes me wonder if that is why I felt so drawn to this place.

The looking tells me that there is much to do, but fills me with overwhelm as I increasingly realise that I don't know what, exactly, or how. In the same caught breath, it calms me as I slow down to notice the details. The old nests thick in the hedges and trees, the birds that flit about and cock their heads curiously at me, the bee under a leaf that I was certain must be dead until I noticed it moving around the next, warmer, day.

I eventually realise that I am studying the land. I am reading books about gardening in hope of learning what to do, but in my increasingly frequent bouts of walking around looking at the space, I am attempting to read the place itself. I seem to be trying to apply on a larger scale what I have learned to do in parts with my houseplants. I only hope I can understand what this whole garden is telling me.

My houseplant collection had seemed enormous in our tiny Oxford home, filling a green wall of space in our living room. Here, they barely fill the conservatory, and I suddenly realise how small each plant is, how many little plant babies I have been caring for. With the backdrop of the garden, it feels clear that I should group them together, give myself fewer, bigger pots to look after if they are to have a chance of survival. There is a lot of ground to tend here.

At least so far they look happier than they usually do in midwinter. The conservatory is cold, but there is far more light than they would have received in our old home, even in front of the double-height, south-facing window, gazing directly onto a neighbour's wall as it did. Here, the plants and I look out onto the garden terraces climbing above us. When I walk to the other end of the house and stand in front of the French doors that lead onto the deck,

I look over the field and wood opposite, can peer through the viaduct to catch a glimpse of the countryside opening up beyond. We are tucked into the landscape here, but the sense of space is far more expansive than our previous scope, where we were hemmed in by walls on all sides, the only views reaching up and away, to the clouds through the skylights.

I look up through the glass lantern roof of the conservatory. After years of staring down dark tunnels looking for the light at the end, I stand here in this winter light-filled room and wonder: what next? Part of me desperately wants to root into this place, and feel at home here, but my whole life, and the lives of generations before me, have been spent in such a constant state of upheaval and flux that I wonder if I can. Empire created the shockwaves that set us adrift; now here I am, on my very own plot of Imperial land. Something has spiralled close to full circle, but I am not sure I know how to pull the threads together into a knot of belonging.

***

One of my paternal great-grandfathers, a man of African descent, lived in a tiny rural village near Trinidad's wild Atlantic shore. The story that has been handed to me is that he was a master craftsman, a carpenter, who went blind. Sightless, he continued to work with seeing hands. Every week he made his way through the tropical rainforest between the village that held his home and the market where he sold his trade, a route that crossed seven streams. Such profound knowing of the land to guide him, sure grounding of his feet in that soil, despite the violent wrench that landed him there to begin with. How receptive both he and that land must have been for such deep trust to grow. Did he feel at home there, held by that place? I think about his story as I wander through my garden, wonder whether my feet will know the crossings of our stream as intimately one day.

A small spot of green on an otherwise bare patch of earth near the stream catches my attention. Unthinkingly, I make my way to it, reaching for my phone to help identify it. But then I realise that without my knowing I have come to recognise it – the plants are young primroses, not yet mature enough to flower. The soil here, unloved as it looks, must be receptive to primrose seeds. I peer more closely at the ground that had seemed so devoid of life a few weeks ago and see the young plants spreading around the back of the beds; I live in primrose habitat.

The primroses' home is cool and moist. They are an indicator species of ancient woodland, of the temperate rainforests which are unique sites on this island that still evoke feelings of magic and wonder. The most complex, rich and threatened habitat in the UK. Child of tropical rainforests, still ravaged by colonial agricultural practice, I am an indicator species of threatened habitat. I look at the primroses migrating out from my feet over this ground and wonder whether we both belong.

My grandmother's oft repeated words come to me. 'You need cooling,' she would say, as she dabbed her bosom and brow with her favourite, astringent Yardley's lavender water sprinkled on embroidered, white cotton handkerchiefs.

'You need cooling': she offered tart concoctions of aloe and mauby as the obscure childhood rashes to which I was prone etched their raw cipher across my inflamed young skin. Words repeated by the acupuncturist I saw on and off during my infertile years, who told me I ran too hot and dry. I didn't understand it, but I knew it. The rashes had largely waned with puberty, but the fire merely retreated within, where it still raged through my organs and constantly threatened to set alight the tinder of my barren life. I take off my gloves and rest my hand, winter skin verging on ash no matter how much I hydrate or moisturise, on the soft pillow of moss that lines the stone edge of the stream next to the seedling primroses. It is cooling.

My daughter comes to find me in the garden as dusk falls. Balancing in her coat and wellies on the stones along the edge of the stream, she sings a song to the tune of the water about the magic and fairies she has found here. She walks over to a more mature clump of primroses nearby and picks a few of the flowers. I stoop to her height and squint, and try to see them through her eyes. From this angle the primroses glow in the gathering dark, luminescent lemon yellow beacons guiding winter moths home. Against them the grey of the garden in the descending winter frost seems to sparkle.

I realise that despite my increasingly regular walks round the garden, looking at the space, I have overlooked the primroses. In my repeated dismissal of their muted presence, I have not looked up whether they have any herbal uses. I sit on the cold rock steps next to them and open my phone. I learn that the plant is edible, the leaves a flavourful salad, the flowers as sweet as they look. But one bit of information in particular catches my eye. For centuries, the primrose has been useful for healing wounds. It can staunch trauma's bleed; it can gently soothe this skin I am in.

Hellebore

THE WINTER NIGHTS are exquisite here. There are no streetlights, and after the perpetual twilight of the city home we had left, the dark feels eerie and profound. I have spent much of my life thinking of night as the absence of light, a darkness to be feared and kept at bay, but here the starry expanse that lies like glittering velvet above us shows me brightness of a different kind. On dry evenings, I wrap myself in layers to insulate against the cold, and go out onto the deck. One evening I stand in the darkness, head tipped back, looking at the stars wheeling above me until I am dizzy. One bright star twinkles, and I try to understand that the flash of light I just saw is a form of time travel, having been made by the burning star millions of years ago, before travelling across the galaxy to meet my eyes. The thought makes me so giddy I have to lower my head, and ground myself in the dark outline of the trees.

The stars sparkle like the frost that forms, an earthly reflection of the clear night skies. I shuffle my feet in their woolly slippers, careful not to slip over on the ice-rink deck, and pull my blanket more closely round my shoulders. The cold air slices my nostrils as I breathe. I sigh an exhale and the only cloud in the crystal sharp night

43

rises before me. I gaze up at the sky hoping to see a shooting star, even though I know it is unlikely. We missed the meteor shower that the almanac indicated in January, but I remember the first time I ever saw one, and stay out in the cold a little longer, in hope.

I had been standing on a wooden deck as slippery as this one, in the back garden of a house we were renting. It was also winter. The children, babies then, had been put to bed, and I was drawn out into the icy night by some instinct I could not name. I stood on that deck in a long dress and my slippers, a little of the warm air of the house tented around my legs, and gazed up at the sky. The Big Dipper, the North Star . . . and then I saw it, almost heard it, a blaze of light fizzing low across the horizon. One spine-tingling moment lit up, and then it was gone, the sky as empty as it had been a minute before, but my chest full of the afterglow of unexpected emotion. As the echo of something magical reverberated, I made a wish; I wished for home.

The winter sky above our new home is magnificent. I stay out in the cold for as long as I can bear, before turning back into the firelit living room, euphoric with awe and cold. I toast my numb fingers and toes before the fire, victims of Raynaud's, and ask myself what I might wish upon a shooting star now. The phrase 'careful what you wish for' comes to mind in my grandmother's voice. It makes me wonder if this home will be everything I desired, or if we will come to regret our leap of faith; the fizzing hurl of our lives into the open blackness of fate.

***

We begin to settle into village life. We meet many friendly new faces, try to remember their names. We accept invitations to come round for coffee, or playdates, extend them in turn. Everything is still new and strange, but we are doing our best to feel at home here.

All this friendly contact with people foreign to us brings consequences. The plague is upon us. Both children get chickenpox, first one, then the other. I am quarantined with fevers and pustules while a helpful network of kind mothers takes whichever child is well enough to school or playgroup, brings them home. I think about how much more impossible this would have felt in our old life, and gratefully succumb to the weight of hot foreheads on my lap, unable to tolerate much beyond too many hours of children's TV.

In my attempts to soothe tender, irritated skin, and the inflamed spirits clothed therein, I remember the infusion of lavender that I tried in a simple poultice last summer when tiny, plump fingers grabbed a hot handle on the stovetop in that fleeting moment between gazes. The poultice was an act of hope more than expectation, using lavender hastily gathered from our neighbour's generous patch. Despite my inexpert application, it healed the burn uncannily well. There is lavender next to the front door here, but when I inspect it to see whether I can repeat my previous cure, I am disappointed. It looks near-dead in the winter wet. So I use the gels and ointments from pharmacy shelves on itchy limbs and hope for the best.

***

When living with my babies surrounded by the constant traffic of cars, buses and ambulances in the busy streets of our neighbourhood in Oxford, I had begun to turn to the plants around me to see if they had any answers to the perennial issue of how I might root into foreign ground. Instinctively, I did not look to the glamorous cultivated garden plants, mostly foreigners welcome for their beauty, transported along colonial trade routes with other commodities and people. Like the now much-maligned buddleia, their welcome was always uncertain, and conditional. Instead, I turned inward, and downward, to the unwanted, insignificant beings in the crevices and cracks underfoot – the city's last wild place. I started to notice the weeds.

The first was a bright blue flower thriving in the petrol fumes of the communal car park. I picked a small posy, and put it in a bottle by my bedside. It had tiny, luminous blue flowers with yellow centres, incongruously named green alkanet for its use as a dye plant. I looked it up, learned that its pretty flowers were edible, and encountered that soon-to-be-familiar gardening term: it was invasive. The word landed with a nauseating thud of recognition. Presumably someone uprooted it from the land to which it had belonged, carried it halfway around the world, and deliberately planted it in this place. Then, when it had adjusted, fulfilled whatever human desire it had been brought to satisfy, escaped its allocated bounds and thrived too well here, it was denigrated as a scourge. My scarred DNA knew that narrative. Most of the major gardening websites seemed to be focused on how I could exterminate this delicate horror. I looked at the flowers next to my bed, yellow centres shining out like eyes meeting mine. This weed and I needed a different outcome.

In trying to think differently about the weeds and outcasts of the garden world, I stumbled upon herbalism. I began to learn about the original sources of the plastic-wrapped, blister-packed pills that colleagues prescribed patients, which sometimes proved so problematic for those who sought my care. I started to notice common herbs on walks, recite their names: dandelion, ground ivy, self-heal, yarrow. Plantain, but not the sweet banana-like fruit to roast or fry that I sought out in the small grocer's down Oxford's multicultural Cowley Road when my palate grew too starved of home. Instead a small rosette of ribbed leaves that lined roadsides and path edges and wherever people tread, and was useful to treat my hay fever.

Trying to find more of these common medicinal herbs around me, often seen as unwanted weeds, I started to venture out into small green spaces that I had never noticed much before. At first I felt awkward and out of place, but the more the plants became familiar to me, the more I began to feel at home in the spaces they occupied.

I sat in the shabby park tucked down a side street next to the hospital eating a sandwich from my handbag on my short lunch break. I took a brief walk along the riverside that ran hidden behind the houses after dropping off my daughter and before dashing to work, breathing in this narrow strip of green and wet before returning to tarmac dust and grey. I visited the unkempt verges, the in-between spaces. No-man's lands; they belonged to these rejected weeds and, outsider that I felt myself, I was comforted to meet them there.

I grew up with my grandmother making a tea for every ailment and I began to make them myself. Remembering them as pungent and unpalatable, I had rolled my eyes at her unschooled superstition and the community's mistrust of modern medicines – White man Obeah – and had only reluctantly consumed the stewed and brewed old wives' tales. I worshipped Science, in its pure and holy white coats, so had not attended to her teachings on muddy herbs, black skin tending brown soil. Anyway, those were all tropical plants and not to be found here. Or they should not have been.

The first time I saw aloe vera water in the supermarket branded the latest superfood, I felt simultaneously proud and ashamed: proud of the knowledge she had somehow held on to despite all the severings, ashamed of my wholesale rejection of her wisdom. I had drunk the bitter aloe water as a child anyway – with a Caribbean grandmother I did not have much choice. I did not buy this expensive, plastic-wrapped, preservative-filled version.

Learning from the urban landscape that surrounded me then, but not trusting it to provide plants fit for consumption, I sought out dried herbs online. I drank green teas for immunity and red ones for my womb, yellow ones that sang of sunshine. I bought pots of unusual herbs to add to the crowd on our patio, deliciously expanded my palate and vocabulary with fresh artemisia, lovage and sweet cicely. I had been introduced to the herbs of Traditional Chinese Medicine in the course of seeking help for infertility, had hesitantly sipped the teas that flowed strange over my Western palate, wondering about appropriation and uncertain of their benefit.

But it was the ones that thrived in the cracks that most intrigued me. Exhaust-blackened and petrol-tainted as they were in the streets that surrounded our small terraced home, there was still a world of potential medicines right beneath my feet. Plants who were healers as I had felt called to become. Plants whose place here was as toxic and uncertain as my own, even if they were indigenous. I felt most drawn to these, strangers in their native land, at home in their unbelonging. We could not be more different, but somehow they felt like kin.

I found myself wanting to get to know the plants of this place inside and out. I wanted to immerse my cells in them, to rebuild my ever-evolving tissues using the very stuff of this place. I wanted to heal my ruptures, to bridge the chasms and join the parts of myself into one connected whole, and I had the tentative idea reaching out from the half-remembered lessons from my grandmother that the plants would help me. But the more time I spent with them, the more I realised that those cityscape plants could not do enough for me.

I needed more than a temporary verge; I no longer wanted to inhabit an uncertain, unstable no-man's land. I wanted to escape those bounds, put down strong roots, and thrive. I wanted my own patch in which to grow healthy and strong. I wanted my own ground: warm and beautiful, safe and loving, permanent and secure. I wanted home.

***

The full moon arrives but I cannot see it for cloud. The storm for which this full February Storm Moon is named rages. Sitting in our hollow in the land, we are sheltered from the worst of it, but from the conservatory I watch the branches of the ash and birches above us whip like blades of grass in the gales, and feel disturbed. The children are wild, like they always are in wild weather. It is as if their cells become as charged as the air around them. I settle yet another squabble and know nothing will calm until the weather does.

Fast on the heels of storm Ciara, Dennis rolls in, broken branches a thick mulch on the garden beds, primrose flowers pinned beneath them. I have never lived so close to the weather in this country, and without the buttress of the city around me, in the face of this winter fury I am uneasy. The wind howls through gaps in the wooden sash windows, and blasts from the poorly blocked up hearth in the middle of the house. I learn all the ways the house can sing, and it sings to me of the cold.

On the first day of the new year, as we finally sat down after tidying up, having waved farewell to the last of the Christmas season's guests, the heating failed. There was the muted boom of an explosion, then the sound of dripping. The boiler had died.

We knew that the heating system needed our attention when we had looked round the house in our early visits, peered into the cupboards and shudderingly shut the doors on the tangle of wires and corroded pipes that lay within. We understood that the mess brushed beneath the carpet of the charming surface was probably why we were able to afford this house. But we hoped that it would last us through to spring, turning valves on and off to adjust the pressure in the system with increasing frequency over the holiday period, the equivalent of putting fingers in our ears and humming loudly to ignore the message the cooling house was telling us. We crossed our fingers and told each other that it would limp on through the rest of the holiday break, and we would contact a heating engineer as soon as tradespeople were open again for the new year. Despite this best-intended denial, we inevitably found ourselves at the start of January with the boiler, the hot water cylinder, the immersion heater and the oil tank all condemned. At least it lasted through the main celebrations, we told ourselves. Cold comfort.

After a week in which I had tried to embrace the health-giving properties of numbing showers and instead developed a worrying cough, I felt cold and dirty. The children ran around in too few layers as ever, unbothered. Oli had the facility to shower at work, but I wore all my jumpers and began a round of unfamiliar leisure centre visits, looking for the most comfortable and convenient place to occasionally wash myself clean. A couple of neighbours learned of our circumstances. There were so many tales of similar occurrences on moving into their cottages that I half wondered if the village was cursed, but as the news spread we were blessed with loans of electric heaters and large urns for boiling water. Kettle baths amused the children greatly; at other times I drove them to my in-laws, grateful to be close enough for us all to receive a semi-regular soak. I stood under their shower and washed my hair and ran the too-hot water long enough to take the sharp wintry edge off my bones.

Along with the heaters came offers to use showers in neighbouring houses. I accepted them all smilingly, then ignored them as the polite but meaningless pleasantries I assumed them to be. One neighbour, our nearest, was insistent. She left a key under a plant pot near her front door and told me I must use it whenever I wanted, that I would not be an inconvenience. But I did not take up her offer until the next time she caught up with me. There was a puzzled edge to her insistence that time, which softened when we sat over cups of coffee and finally understood each other.

She really meant this offer, it was honestly no inconvenience to the bustle of her household, there was room enough in her home for me to come and go as I needed. I was genuinely baffled by this kindness, unable to conceive of a person inviting the strange Black woman who moved in next door a few weeks ago to shower in their home whenever she fancied. She remembered how she felt moving here from a big city twenty years ago, how difficult it was at first to trust in the offers of community. I began to realise how I had been scarred by more than twenty years with my guard smilingly up. The same protective layer necessary to ward off hurt now impeding my ability to accept help, to access connection. I went home, soul warmed through, body and heart washed clean, and with the disturbed sense that my long-held ideas of belonging and acceptance must shift. I began to sense how vulnerable it might feel to be truly embraced.

It took a while for the thick stone walls to shed the last of the heat that was held within them, but soon I had chilled so profoundly that I became resigned to it. While waiting for the new, renewable heating system that we had decided on to be installed – the reserve of money that we had set aside for redecorating the house covering the quote so exactly that it seemed meant to be – we fell into a routine of borrowed baths and showers, borrowed electric heaters, a lot of baking and cups of tea. Many jumpers, extra woollen socks. I spent the time that I was at home in the kitchen, cooking. It seemed the warmest thing, and I was always hungry. A friend sent a link to an article about the exhausting effects of cold and pain on the body, and suddenly things made sense. The cold was taking a toll. A toll which the other members of my family, all English born and bred, seemed not to be experiencing. I had lived here for half my life, and adjusted well enough, but still my bones needed heat in a way that others' did not.

***

I head out in the light of day, as even in the February sunshine it feels warmer than it does in the house. It is a beautiful winter morning. The day dawns with a clear blue sky, and the bare tree tops of the wood glow orange in the early sunlight, catching the beams streaming low from the horizon. Columns of steam rise from neighbouring houses as inhabitants wake and start their days, the largest, just visible through the viaduct arches, coming from the farm down the valley. Everything is etched in ice after a cloudless night, and the field around our house, still in shadow, gleams silver before the sun rises higher and the frost melts in its warm caress. My breath is made visible around me as I bundle tight fists in my pockets to keep my fingers warm.

Turning left out the front door, as I go down the steps to the driveway, I notice that snowdrops are here. Delightful clumps of their dark green, narrow leaves gather in the shelter beneath a tree that sits next to what would have been the cottage's original front door. A few early white bells dangling over the leaves let me know who they are, and I pause to admire them. Bright spots in this dark corner, they seem to shine on some pink and white flowers rising from the ground around them. Hellebores in their dozens.

Delighted by this discovery, I lift their bowed heads to more closely admire their pretty, freckled faces. My husband and I were given a hellebore by his grandmother many years ago, when we first got together. Flowering for my February birthday, they had become my favourite winter flowers. We placed the hellebore we were given in a pot, and tended it for more than a decade, but, in a metaphor too rich for our relationship to our old home, it became increasingly potbound and unhappy. This spring there is no sign of the potted plant, the blue vessel in which it travelled over from our old house to sit scattered among the rest of our portable garden on the patio still empty. Discovering this bountiful patch of hellebores feels like a blessing.

Inspired by their presence, I wander around the rest of the garden to see if there are any others that I have not yet noticed. As the frozen gravel crunches under my feet, I notice dotted beneath the beech hedge still holding onto its brown autumn leaves the feathery green of what seems to be another type of hellebore. I crunch around the conservatory to the other side of the house, and there are more beneath the bamboo, and rising among the ground cover in the ever-shaded bed that runs behind the house. Despite the morning frost these ones stand tall, with large, leathery, highly divided leaves, in places that look strange to me. I cannot tell if they are weeds or deliberately planted. Looking at them more closely, I realise that the clusters of pale green, upside-down cups held on thick, contorted stems above the leaves are flowers. I am not sure how much I like the look of them, uncertain whether I find these tougher-looking hellebores ugly. With cold fingertips I look them up on my phone, and learn that these are the stinking hellebore, an indigenous plant, and one that looks comfortably at home in my garden.

My eye falls on an article that mentions their herbal uses. Having an idea that they were toxic, I am surprised by this, but pleased. With the garden feeling unknowable in its depths of winter slumber, I have no fully formed vision for how I might like to shape it, but the beginning of an idea of a space made beautiful with plants that all have some edible or herbal purpose has taken seed. I cannot yet articulate the feeling that is growing with our time here, as I tentatively begin to spend a bit more time in the garden, but I have a sense that our home needs to be a place that will nourish and heal us. I click on the article, and read that some hellebores were once considered powerful medicines for various ailments, but like all the strongest medicine are also extremely poisonous, so their use has fallen out of favour. That which might heal can easily kill, if used inexpertly, or taken to extremes.

I am not sure about the native hellebores, but as I walk around led by the sound of the stream loudly rushing through its stone channels, I am certain about my love for the garden's water. When we began thinking about choosing a home for ourselves, and a place to settle into, I was drawn to the idea of living by the sea. Being close to water has always been a comfort for me, and one of the things I found most challenging about living in Oxford was how landlocked it felt, and how far it was to get to the sea. In the end a coastal location did not present itself as an option, but this spring running perpetually, cheerfully, through this garden seemed to speak to the water-loving part of me, and reassure me that having it might be enough to slake my thirst.

I love the water here. In this bare, empty, late winter garden, the sound of the stream is the garden's most prominent feature. Every morning I open the conservatory doors and follow the clouds of my breath across the stone to the small green cascading pools on the other side of the patio. Every morning it strikes me as magical that this cold, clear water appears here of its own volition, runs through this space. Water is necessary for life; we barrel on our blue planet through a galaxy full of millions of arid rocks on which no life or water can be found. And yet it has chosen this place. The water sings to me, and I feel the dry, drought-ridden places within me begin to be quenched. There is something about the stream that feels special here, its constant burble comforting, and watercolouring my dreams.

I had started dreaming of the water in the garden shortly after our arrival, of sitting by it, listening to its song. On waking I could never remember what it had been trying to tell me. When the heating failed and the engineer assessing our system warned of a possible oil leak into the stream, his warning felt like déjà vu. It felt like remembering.

***

Late winter squalls weep over us for days. Their passing leaves the valley flooded, our bins floating, the bottom garden a new lake. It is wet here, the village named for its many natural sources of water that bubble year-round from deep in the earth. But I walk through our neighbour's sodden field in a break in the rain and see dozens of new springs welling beneath my feet. Villagers frown, speak tightly about never having seen it so wet, the rainfall extreme even for this lush place. Every morning we peer into our cellar, where a now-worryingly ad hoc system of a pump on a timer keeps the small pool of water that gathers in the carved out foundations at bay. This house has stood for hundreds of years, we hope it should stand for many hundreds more, but the intensity of weather it was built for is changing. This easy source of fresh water, once a blessing for all the humans who have settled in this spot for millennia, may prove as the climate changes to be our future curse.

The flooding in our garden abates in time for my birthday. This is a relief, because the heavy machinery that is to finally restore warmth to us has just arrived. The flat lawn next to the driveway at the foot of our garden is destined to hold a pair of 130-metre-deep bore-holes, down which a system of pipes will run. Fluid piped down to the bottom of the boreholes will be warmed by a degree or two relative to the temperature at the surface.

This small difference will be converted by a heat pump, acting like a sort of reverse refrigerator, into water warm enough to run through our underfloor heating and thaw me again. The underfloor heating that somewhat surprisingly runs throughout our rather rustic home happens to be perfectly compatible with a ground source system, and the geography of our ground water is also a boon. Water running through pipes under every surface of the house, heated by the water in our heavy soil many metres underground, to protect the water that sings in our stream from future poisoning. It feels fated. As does the booming of the rig making its way deep into the earth's core to ring in my thirty-ninth year.

***

Carnival arrives, not that my home in England knows about it. Because of our house move, we are sitting the jubilant festivities out at a distance. It is my favourite time of year at home, the liberation of losing yourself in a wild two-day street party a fitting answer still to its origins in slave rebellions and uprisings. When we do make it back for Carnival, I do everything. Attend a series of the huge fetes, where thousands party, drink and dance to the frenetic beats of modern soca. I visit the steel panyards to lime with friends while we hear the bands practise, before hitting the Savannah for the grand competition of the orchestras on Carnival weekend. I attend the calypso tents, where swaggering bards perform their calypsoes laced with socio-political commentary and double entendre. I wake before dawn on Carnival Monday morning, *jour ouvert*, join with revellers in the streets for J'Ouvert. We cover our bodies with mud, and paint, and oil for dutty mas, the riddim of steel reverberating through my hips as we chip, chip, chip, feet rhythmically meeting street. We embrace the black devils the old massas portrayed us to be, and then I go to the beach when the rising sun banishes us and am cleansed in the island's waters, sleep like the dead on the sand. Carnival Tuesday, the grand finale, I rise again, phoenix in beads and feathers, fire in my waist as I wine like a godless goddess.

It is the greatest festival on earth, and instead I am here, tinny sounds and pixelated images on my laptop screen while I make the children pancakes.

I have already vowed that next year will be my year, an appropriate setting for my fortieth birthday celebrations. Meanwhile, a lovely surprise. I have had a text from an old Trinidadian school friend, on a once-in-a-lifetime tour of Europe for her fortieth. She has arrived on the English leg of her travels, which means that she can come and visit us.

We walk around the garden in the winter sunshine, sit at the top of it with glasses of wine, looking down at the garden and house below. We cannot stay long; despite her tour she is too tropical still to tolerate much sitting outdoors in late winter. But it is briefly delightful, one of my oldest friends in my new garden. 'Wow,' she says, gazing at the view below us. She turns to look at me, and I feel her studying me in my new setting. I look away, at the garden cascading beneath us, nervous of her appraisal. Eventually she sighs as she takes another sip of wine. 'This suits you. You look really happy here. I'm so glad I got to see it.'

The next day, I go to the garden centre. I am looking for shade-tolerant plants, to enhance the area where I sat with my friend at the top of the garden, near the small gate through which we enter and exit on our way to the rest of the village. It is clear that this gate was not much used by the previous owners, and with no formal path across the grass we are quickly wearing a muddy track. The path needs proper thinking about, but right now I want something to quickly make this space feel better.

I put some sky blue brunnera and pretty pink pulmonaria in my trolley. I wander slowly round the aisles, dithering over what else to buy, when I spot them. A glorious display of hellebores. With their glossy green leaves, and speckled pink and white faces, I have decided that the plants in my current favourite bed in the garden are *Helleborus orientalis*. Here in the garden centre, I am drawn to the boldest plants, dark purple flowers with yellow stamens and leaf veining almost neon in contrast. They stand majestic, flamboyant masqueraders among the duller winter plants around them. They seem to be a different, hybrid species of hellebore, but thrive in the same conditions as their native cousins. This reassures me that they will feel welcome in our space. I put the most magnificent in my trolley.

***

We watch the work progressing on the new heating system with fascination, but my inability to do anything to influence the much-hoped-for success of this huge-scale project makes me nervous. As does watching the necessary destruction of the beautiful part of our garden which formed my first, captivating impression of this place.

To counter my helpless anxiety, I decide to take control of the rest of the garden. Perversely, this also seems to mean destroying it. To one side of the conservatory lies a bank of enormous bamboo, planted by the previous owners in order to screen where a trampoline once stood. It has since grown many feet tall, and far escaped the confines of its planting, casting what should be a light-filled garden room into creaking shadow. One day, I stand in the middle of this small bamboo grove. I feel cocooned in green, and think of the Bamboo Cathedral of my childhood home, a similar space on a much larger scale on the northwest peninsula of Trinidad. There, the bamboo grew as high as the trees with which it competed, stalks thick and sturdy as limbs, strong enough to be cut and used as scaffolding. The arching stalks which met high overhead looked like the soaring nave of a cathedral, and the space within felt sacred.

We pilgrimaged there every time we went swimming at Macqueripe Bay, a deep, perfect horseshoe near the northwesterly tip of Trinidad. It was my great-uncle's favourite swimming spot. He was a GP, and went for a swim early every morning before opening up his practice to his devoted patients. He lived in a big house on a hill overlooking Port of Spain, and was always relaxed, and his lifestyle had charmed me into the idea of medicine. My experience felt very distant from his lush one.

I look up local businesses on the parish website and hire a gardener. He agrees to clear away the patch of bamboo, then continue to work regularly with me to give the garden a winter overhaul, and to guide me in the tasks I should be doing over the coming months. I feel reassured by the idea of his guidance and help. Suddenly the garden seems a less daunting place.

Rapidly giving up on the idea of digging out the bamboo, which we will find months later emerging from the path beside the house, spreading under our feet with a will entirely its own, the gardener settles for cutting it back. It takes him several days, hacking through the thick stems, a growing patch of vicious stubble in his wake. At the end of his first week, standing in the new light next to the razed beds, he tells me that he will not be returning. The garden has overwhelmed him, presumably a trained professional, and I, amateur, am left to figure out the butchered site on my own.

After he has gone, I stare at the steep bank covered in fallen bamboo leaves, a sense of panic rising in my chest as sharply as the land above me. Unlike the other garden terraces, this area seems unfinished, jutting half-formed steps made of tarred old railway sleepers rather than gently curving stone. The bed bristles sharply upward to meet the thorny boundary hedge; this is hostile ground. My skin prickles as I feel all the desperate anxiety of trying to navigate the knife edge of surviving in this unwelcoming land spread over me. I have never owned a garden in this country before, and standing here, just behind the house, where I can see all the rest of it rising above and falling away from me, the conviction that I have taken on more than I should, more than I am allowed, and can manage, threatens to overtake me. He has abandoned me in this space; without guidance I cannot trust that I will know what to do. I feel an old anxiety well from somewhere deep within me. It is familiar, as if it has been written into my cells, waiting for this moment to make itself fully known. It tells me that I will destroy this place, that my dark ignorance can only mar this landscape's pure beauty.

A shuddering breath conjures another garden. The first home of our very own, bought by my parents when I was a teenager. My eyes blur as I seem to see it before me: its blocky structure, with a green-painted, galvanised sheet roof, half standing on stilts in the middle of a steeply sloping plot, a bank rising behind dry with bamboo leaf litter. And a garden to be shaped by us three, amateurs.

That home was my first introduction to gardening in a deliberate sense, and the space eventually grew through our enthusiasm and failures into one I remember as a beautiful, magical place. Our little paradise; our Garden of Eden. The vision rises as if to quell my growing overwhelm. I try to focus on the scene in front of me, the eerie echoes of that childhood Trinidadian home overlying this English one so densely that for a moment I no longer know where I am. Then a robin flutters in, curiously inspecting what has been revealed in the bamboo's shorn stalks. I take a breath, and my cue from the bird, and look at the slope with clear eyes. This is a landscape ragged with possibility, and one I have seen before. I am free to create what I like here. I am liberated, and I am terrified.

Awakening

Blackcurrant

THE THICK STILLNESS is eerie, tangible. The horses clip-clopping past our front windows are gone. No cars accelerate up the hill. The sky is a brilliant, clear, windless blue, scrubbed clean of cloud. The colour of fresh starts and hopeful beginnings, fitting, as we marked the spring equinox only days ago. We are on the ascent, riding the wave of the year to its peak. I push open the French doors and wade through the dense quiet to stand on the deck; even the birds in the wood opposite seem muted, their song indistinct.

One sound slices through the uncanny calm. Beneath me, a loud, slow drip falls from the cracked sewage pipe that crosses the open trench of the abandoned groundworks. The landscape has shifted beneath our feet and the strain was too much for the old, badly laid pipe to bear. I cannot yet see any shit in the murky water below, but I know it is there.

Lockdown.

Looking at the still-naked trees, I glimpse that barely-there shimmer of bright green, of new leaves nearly ready to burst from their winter buds. The thought of spring is a comfort. The heavy quiet enfolds me, and my shoulders drop under the weighty hug of reprieve. I sigh into the clear balm of this belated certainty. Stay at home.

Along with millions of other British people, we are locked into our houses, the long overdue, utterly shambolic government response to the soaring crisis of lives lost to this new virus, this Covid-19. Lockdown comes as a long exhalation of relief – at least for those of us for whom the idea of home is a sanctuary. For us lucky ones, it is clear what to do now. The anxiety of carrying on as normal, the fear of every interaction potentially being a chain reaction of death, has been wiped clean in one direct mandate at last. Stay at home.

The official instruction has been that this unheard-of state will last for three weeks initially, and then we will reassess. Three weeks. Twenty-one days. It is not so long to bear. The drip of sewage into the mire beneath me carries with it the weighty stink that, given the delays, given what feels like a horrible lack of urgency and care in our leaders' response to this public health crisis, we might be living through this dread for a lot longer than a few weeks.

I am not yet ready to fully hold the implications of that reality I can sense stretched out before me, like the churned up landscape below. The aftershocks of our own upheaval to land in this new home have barely settled; tremors rumble through us still. Now the entire world as we have known it has been thrown into disarray, our collective sense of normality completely uprooted. I need the deep breath of relief, the lightness of optimism, to ground me in the still of this strange moment. The stream babbling over mossy boulders along the edge of the mud strikes a cheerful note against the becalmed morning. I raise my eyes to the blue sky above the trees and inhale deeply the fresh stench of deluded hope.

***

As I do every day now, I go out into the garden. Restricted to our own space apart from one short walk a day, I have no choice but to begin to intimately acquaint myself with it. This place to which I had felt so clearly called on our first visit, which had evoked such a startling reminder of the tropical garden I had once thought of as home, but which had repulsed me all winter with its ugly reflections of my inner landscape. Lucky enough to have a garden, but trapped in it by lockdown, I cannot avoid it any more.

We are nearing the end of March, and I look at the exposed dirt of the main ornamental beds in the terraces above the house, hoping and expecting to see bulbs emerging, those tellers of the spring. I can see tantalising buds forming on the slowly waking arms of as-yet-unknown shrubs, but at ground level, apart from the small bed next to the driveway covered in gone-over snowdrops and still-flowering hellebores, and the nodding clumps of primroses that line the stream, there is still little of interest. Our garden is at the lowest point in the village, and the temperature drops noticeably as you descend into it. Frost lingers longer in the shade here, and I wonder if this is the cause of its late wakening. A bright handful of daffodils under the quince tree at the very bottom of the garden serves only to heighten the sense of bare earth around it.

I am starving for colour and beauty and joy to leaven not only the usual end of winter blues, but the weighty dread of the strange and anxious days we find ourselves in. I resolve to buy and plant many more early-flowering bulbs this autumn.

As I wander the garden with a coffee, searching for inspiration as to how to occupy the children I am now homeschooling, a flash of rich berry pink at the back of a bed near to the beech hedge catches my eye. I make my way along the stones that line the stream to get a closer look. Tucked away right at the back of the border is what an internet search, when I temporarily retreat to the house and its wi-fi, tells me is a blackcurrant – *Ribes sanguineum*. Each cluster of unfurling flowers hangs like a ripple of blood-red droplets suspended on not-yet-fully-leafed branches, before the petals of each flower curl open to reveal a heart of softer pink. Against the dull quiet of the yet-too-still garden, it beats a spectacular refrain. I nearly gasp to see it.

I pluck one of its leaves, irresistibly drawn by the ruby of its flowers to touch like a child. The leaf is cashmere soft; I rub it between my fingers and against my cheek and smell its berry tang. It is delicious, and familiar, and, in the way that smells so often do, in memory immediately transports me someplace else. I am back at home, in our cobalt blue kitchen with the terracotta floor tiles, helping my mother make the jewel-pink, floral-berry cordial that we call sorrel. Not the lemon-flavoured, leafy herb I some-times see in English garden centres, but the *Hibiscus sabdariffa*, whose deep pink calyx is picked and dried and used to make a spiced cordial, often drunk at Christmas.

I can almost hear the bubbling of the ruby liquid in the pot, the sorrel sepals simmering with cinnamon and cloves to make the distinctive drink. My mouth watering at the memory, I pick a few more leaves of this blackcurrant for a tea that will smell of the childhood home I am now profoundly cut off from. I cannot yet bear to let myself feel the full meaning of being so split off from my family and the beloved landscape of my birth. We settled in this home with the firm understanding that there would be regular trips back to the Trinidadian landscape that feeds a vital part of me. The idea of two homes has always kept me split, but the door to one is firmly shut now. I do not know what that will do to me. All I can let myself feel in this moment is hope that the separation will not be for too long, and gratitude that the garden has given me this plant, this surprising blackcurrant with its reminder that offers some small tonic for my soul.

Inside, the children are creating a racket. Strung more tightly than my anxious nerves, they twang and ping off each other, an innocent distillation of all our confused feelings in the face of this uncertain situation that none of us fully understands. My senses will need a stronger tonic than blackcurrant-leaf tea to get us through this.

***

I came here looking for home and now I would have to face it. It was unavoidable; I could not leave.

We are all in this together, I read, as Oli dons his Lycra and gets ready to cycle off to the hospital, leaving me to yet another long day at home alone with the children. I sit on the edge of our bed, glumly scrolling through the day's horrific headlines on the news app on my phone as he pulls on the long black leggings with the faintly ridiculous padded seat and rests the elastic suspenders on his shoulders. He zips his forest green jacket over the top as the children tumble into the room. 'Daddy, Daddy, don't go to work. Stay home.' It is our new morning routine. He will peel the children off after giving them long hugs, give me that look that I know too well – the one that says I love you, I don't want to go, I have no choice – before squeezing me tightly, his face buried in my hair, my cheek pressed against his chest. He will put on his muddy bike shoes and mud-spattered, neon yellow, hi-vis waterproof and helmet in the porch, then clip himself into the bike at the top of the garden. He must remember to slow down as he passes our front gate, to allow the children to wave him off, beseech him one last time to get off his bike and stay. Then they cry, and I must swallow my own tears and comfort them.

The routine is mirrored on his return. We know colleagues who have isolated themselves from their families to contain the risks of transmitting this dreadful new disease, but we cling to each other. Neither of us could face this completely alone. And so he strips himself of scrubs and clogs in the changing rooms at the hospital, and washes himself clean, hands always slightly raw now, before donning his Lycra for the return journey. We fold back into the home of each other with exhausted relief every evening.

The children have been at home with me for an extra week already, as we made the decision to pull them out of school days before lockdown was finally announced. I learn later that a significant proportion of parents at the school with the capacity to make space for childcare in their lives, from all backgrounds, did so a week ahead of the government announcement, when the only right course of action seemed too obvious to ignore. The school had already provided us with some materials and a hastily constructed sketch of a learning plan, the teachers gazing clear-eyed and thin-lipped at what was coming when our leaders would not. We stayed at home, but it was official now, and the collective pause reverberated through the stillness around.

I ache. With relief that I can be at home with my children, when they so need me. That I can be the stable presence at home for my husband, when all at work is thrown into life-threatening anxiety. I am profoundly grateful for this seemingly divine timing that has allowed me to be here. In this bigger house, unheated as it still is, with this wonderful garden that gives us so much extra space, in this beautiful countryside. In this capacity where I can be dedicated to my children, and all of our well-being. I know how lucky I am to have the space to even attempt to be our solid ground.

At first, we treat it like a holiday. It is novel, and for the children on some level quite delightful to be at home with Mummy. The exercises sent home from school are rather fun. With my oldest child just in reception – and, as I later learn, the national curriculum temporarily suspended – it is all about encouraging any kind of learning through play. This is an approach that I have longed to embrace, but have never had enough time for.

73

All we have now is time, and so we devote ourselves to it. We bake, we craft, we cook together. We do phonics videos, and I learn the proper English enunciation of certain sounds for the first time. Hear, hair, here, heir – all homophones in my native dialect, but I finally understand the peals of laughter that used to accompany my request to drink a bear as a student. I suppress my desire to drink through the lessons, and we practise letters, and do maths in the kitchen. We have endless hours of story time, snuggled together reading books on the sofa, and learn to sew. It is all awfully wholesome.

This retro spirit of cheerful stoicism seems to have seeped into us all. Friends talk in upbeat video calls about Zoom quizzes. We wave from the deck to a steady stream of fellow villagers walking through the field below on their daily exercise. The exchanges follow the same formula: aren't we lucky to be here, we shall have to make the best of it. Alongside tallies of deaths and scenes of chaos in hospitals, tales of our collective, chipper 'Blitz spirit' dominate the news. It is strangely wonderful, until it is wholly awful.

I start to feel desperate to do something more, something other than the relentlessly unending domestic. I had never envied my husband his career path before, seeing all too clearly how much better I was suited to mine. But in these moments of holding my children more closely, more intensely, more in isolation than I had ever done before, I burned with envy for his ability to leave us all behind. I longed to rise above the class Zoom calls, phonics and maths videos, the endless rounds of meals and snacks. I wanted to wipe myself clean from the muddy knees, snotty noses, messy hands and dirty bottoms. Some days on our daily walk through the woods and fields around our home, I wanted to run off to the crest of the highest hill I could see, and scream.

I feel my jaunty construction of stoicism collapsing upon itself.

And not only mine. Empty supermarket shelves, panic-buying of pasta and flour speak of the dread behind the collective determined front. The mass hoarding of loo roll becomes a phenomenon less bizarre in the face of the question lurking at the back of all our minds – how will we possibly cope with all this shit?

***

The days repeat themselves. My thoughts have been spun off-course by the surreal events and circle themselves in an unending loop. Time melts into distorted form.

I read articles linking the emergence of new strains of the virus to the climate crisis. Others reporting that our leaders were made aware of this, and of the shoddy state of our national pandemic preparation plans, and ignored it. This neglect of human life – this abandonment of care, ethics and plain old common sense at the altar of profit – feels all too familiar. It is a tale carved into my bones, gouged from the flesh of those who made me. I can see no light at the end of this febrile tunnel.

Calls for retired and out-of-practice doctors to return to the NHS begin to appear in the newspapers. The GMC licence that I gave up a few short months ago is reinstated under emergency terms. Hearing horror stories from friends at the viral epicentre in London, I contact my local hospital and offer to provide emotional support to staff. The crashing peaks of illness crushing the London hospitals have not reached us with the same force yet, so I am thanked, and held in reserve. While making cheese-and-chutney sandwiches for the children's lunch, I speak on the phone cradled between my neck and shoulder to people recruiting for national medical therapeutic services. What is being sought is ad hoc care flexible enough to work around the ever-changing hospital rotas, where doctors are shunted and shifted from the specialties of their choice to overflowing wards of breathless need. Therapists who can shift with the doctors, to make themselves available at all times, are required. I understand this, and understand that I cannot provide it.

With my husband's workspace ever shifting beneath his feet, I am needed at home with our children. The massive change in their lives when we moved house was upheaval enough; I can sense their deep disturbance now at the horror that surrounds us all. I feel that to send my son into the profound unknown of the essential childcare provided by school while I also return to work would be an abandonment that I might never be able to repair. And I cannot imagine what arrangement I would have to cobble together for my daughter, still too young for school. Family support is unavailable. For all my desire to fully protect the children from the terror in the world, we cannot – their grandfather lies in the most vulnerable category, immunosuppressed, ongoing chemotherapy temporarily delayed. The unspoken fear is that to have contact with us, exposed as Oli is to the viral frontline on a daily basis, would be lethal to them. My sensitive son understands without explanation that his beloved grandfather might die.

And so we go on. Oli sallies forth to the hospital, armed with the flimsiest of protection. As one of the youngest and fittest consultants on his team, the burden has fallen on him – he begins what will become a year of being the inpatient consultant primarily responsible for face-to-face care for his specialty. I choke with the terror of his catching this virus, every story of a healthcare worker who has died in service squeezing the vice on my heart tighter. But though there are mornings the lump in my throat makes it too hard to eat before I bid him farewell, I cannot burden him with my fear. I must hold him, with a sense that by doing so I am holding up all of his patients too. I am holding the world for my children. With what little reserve I have left, I must hold myself, to hold it all together.

***

I volley between extremes of resentment and relief, de-stabilised by the maddening uncertainty of the times. Intuitively sensing that the only thing that can ground my rapidly shifting internal landscape is the relative constancy of the external one, and perhaps drawn by the irresistible call of early spring, I relinquish my fight against the waves of detritus washing up on counter tops and floors and tables in the cold house, and fully abandon myself to the outdoors. The children invariably tumble after me.

In the sudden absence of our normal routines, days

begin to blend into an amorphous, repetitive, unsettling mass. With nothing else to guide me, I turn to the structure given by the garden. In a human world that no longer quite makes sense, nature still does. It is spring, and so we must sow things. With a new garden to tend, and dreaming of spring warmth in our freezing house, over winter I had bought packets and packets of seeds, delightedly poring over catalogues online. The joy of thumbnail images of potential summer beauty! They were lushly irresistible. And as lockdown leads to seeds rapidly selling out online, I am grateful for the prescient packets stuffed into the tin on the kitchen worktop, to which we are drawn again, and again. Microdoses of hope.

To combat the online lack of seeds, a large plastic box of them begins going around the village, from one doorstep to the next. People put extras in, help themselves to ones they need. Someone mentions somewhere that the new arrivals to the village seem to quite like gardening, and the box appears by our door, with a note asking me to help myself. I am overwhelmed by the choice of seeds when I look in, by the generosity of so many potential plants. I try to put in more packets of seeds than I take, before passing it on to the next house.

On the circular patio behind the kitchen the children and I scoop peat-free compost from sacks heaped in an untidy pile into small plant pots. Their little fingers push fat broad beans beneath the surface before watering them unevenly – some merely sprinkled, some drowning. My three-year-old daughter soon loses interest and wanders off to make more interesting potions of her own, but my son is intent to keep going. What shall we sow next?

Over these first days and weeks of lockdown we rifle through our packets and sow them all, in trays and rows of pots, then old fruit punnets, loo rolls, shallow cardboard boxes that come with the online deliveries. Some seeds are fine as dust, impossible for my clumsy amateur fingers to handle individually, and get scattered unevenly. Some are crescent-shaped, or like tiny stars. We marvel over the shapes and colours every time a new packet is opened, all three of us fingering these minuscule marvels that house entire new plants.

Soon, the floor and windowsills and most of the tabletop of the conservatory which doubles as our dining room is covered in spilled compost and precariously balanced trays. We watch for germination obsessively, checking up on seemingly lifeless containers dozens of times a day. And then, the burst of joy at the first sign of life emerging. Noses almost pressed against each seed case cracked, the radicle pushing its way into the compost, delicate hairs fanning out around each tiny root. The glee when the seed leaves manage to push off their hard shell and spread themselves wide, reaching for the light. With each new seedling grows a fresh source of dopamine, each tiny plant lifting the dense weight of pervasive anxiety as it unfurls. It is addictive, and I am hooked.

It is the most ordinary thing, but pushing my dirt-stained fingers into the soil and summoning forth green slivers of new life feels magical. I hum with creative life force, a world apart from the depleted, deadened feelings of a few months ago.

We sow things for beauty – I finally open that packet of the evocatively named cosmos that so entranced me at

Christmas – but, urged on by the children, we especially sow things for food. I have never done this before but, along with what feels like half the nation, we are sowing the bounty of summer, trying to future-proof our precarious lives. The collective urge feels instinctive – primal – a long-forgotten ritual reawakening in all our flesh to ground us in these uncertain times. In our great hunger for the stability that we have suddenly become starved of, we are trying to nourish ourselves.

***

We learn that the farmers' market in the centre of Bath, where we had come to do the bulk of our shopping, is still running. I grew up with the weekly farmers' market, my father making the pilgrimage early every Saturday morning to the main market on the outskirts of Port of Spain, his pile of empty shopping bags ready to be filled with produce. I was not fond of the early start, but would sometimes be dragged along to accompany him.

He approached it like a ritual. Parking in the same spot on a street nearby, where he knew he would get a place, rather than risking the busy market car park. Then winding his way in a familiar pattern between the vendors he had come to know. This one for tomatoes and okras, but never pumpkin. Bundles of green seasoning, chadon beni and callaloo from over there. Plantains, paw-paws, carailli, christophene, pigeon peas, green fig, breadfruit, watermelon. Sometimes cured and salted pigtail. Mangoes when they were in season, and the tiny, intensely sweet bananas with skin so tender they could never be exported.

By the time we had made our round of the market, several hours later, our bags would be laden.

It took us so long to pick our way round the bright colours, loud noises and pungent smells of the market because my father knew all the vendors' names, their whole families. When I went with him, we would often be carrying home extra treasures, freely given 'for Farrell nice young lady daughter'. But he too would often be carrying things to pass on and share, gifts for this one's birthday, or that one's son's graduation. The market was no simple replacement for a supermarket: a place selling the commodity of food. It was much more than that. It was a place of community, and kinship. The food exchanged was more akin to sharing meals with family and friends. So much more than the body was nourished.

It amused me to see my father replicate these relationships in our local supermarket when he and my mother came to visit us when our babies were born. They would take the children out to the nearby playground, and stop in the supermarket on the way home, stocking up on supplies for their Airbnb kitchen, as our tiny home was far too small to hold us all. Soon, my father knew all the regular cashiers' names, and their stories. They would greet me and ask about him when my parents had returned to Trinidad and I ventured round the supermarket alone with my children, whom they had all come to know.

And so it felt natural to go round the Saturday farmers' market when one opened on the main street near our house in Oxford. By the time we left, we had known some of the traders for more than ten years. It had become our weekly ritual, our place of community. Finding the farmers' market in Bath was one of our first acts on arrival.

One Saturday morning a few weeks into lockdown we venture out to the market. We walk past the long queue snaking through the car park to the nearby supermarket's doors, and breathe a deep sigh of relief at the already familiar farmers and growers standing at the too-quiet market stalls, face coverings not enough to mask their anxiety. Here, there are no shortages – fresh produce spills out of crates and lies heaped on tables as usual. In fact more than usual, as the market is quieter than normal. I watch the lines of people walking past, going to wait for their turn in the supermarket, and am struck by how mad it feels, how profoundly split off we seem from the nour-ishment we all seek.

Oli and I get all that we need, and some extra treats too: chocolate frogs for the children, and homemade take-away for ready meals when we are too tired to cook from scratch. We learn in rushed, distant exchanges about other helpful sources of necessities. A bakery in town turned grocer, to which we go next and collect not only freshly baked sourdough but also bags of otherwise scarce and precious flour. This one organic, heritage grain, locally grown and milled, and readily available.

Colonial plantations were the birthplace of modern industrial agriculture, chaining sustainable smallholding practices whose primary aim was feeding communities into large-scale farming with the aim of extracting maximum profit from land, plants, animals, human bodies. As the plantation offspring of big, corporate, unwieldy supply chains collapse in the face of sudden viral crisis, the small, personal, nimble growers and farmers are deftly doing all they can to keep the rest of us fed.

And where supermarket home deliveries have become impossible, when my youngest develops a fever, and we are completely restricted to home, the village shop brings us fruit and veg, and the local baker drops off a crate of bread, and some free cake for the poorly little one, on his way home.

**

The garden has some veg beds created by the previous owners. I like that I cannot see them when sat in the conservatory; the ornamental beds have distressed me with their emptiness, but something about this site of growing my own food feels even more repulsive to me. It makes no logical sense. It is the natural next step in the relationship we have nurtured with our food, and I should be delighted at finally having the opportunity and space to grow the things I eat. I have been lit up with more joy than I could have imagined in the dense anxiety of these days through sowing seeds of plants to eat. And yet a heavy reluctance sits in me.

The veg beds are bare, and contain the same heavy soil that weighed on me all winter. On one of his days off, Oli walks up the shale-lined steps that run between the beds, tries to fork some of the ground over, as he learned from watching his parents manage their veg beds. It sits in unyielding clods. But I have been reading about no-dig gardening. In my last, urgent trips to the garden centres when a lockdown of some kind seemed imminent, I had bought bags of compost. I haul one up the many steps to the veg beds now and spread it out. The rich black soil suddenly looks hopeful.

My nearest neighbour, still taking care of us from a safe distance, texts to ask if we would like some spare bean plants – she has sowed too many. As our own beans have not yet germinated, I enthusiastically accept, and to my surprise find more than a dozen sturdy small broad bean plants outside the back door later that day, accompanied by about half a dozen foxgloves. She grows foxgloves from seed for her shaded woodland bank that rises above both our houses, and remembered that I had mentioned in one walk round her garden that they were Oli's favourite plant. He tends to hearts, and foxgloves contain the precursor to the powerful heart medicine digoxin, which helps to correct irregular heart rhythms. My own heart swells when I see the seedlings, beats more strongly. I feel how we are held in hers.

This determines the day's activity. We are planting out the beans that our neighbour has been sowing in her greenhouse to supply us all. Normally delighted to get stuck in to planting in the garden, I am struck again by how resistant I feel to doing the work of growing food. It brings to mind my grandmother's injunctions about not getting dirty when playing outside. I spent a huge amount of time outdoors as a child, either playing alone in our yard, or running off to play games with friends down the street. Most of my memories of the first ten years of my life are of the outdoors, in dens under hedges, or in the crawl spaces underneath the traditional Trinidadian houses raised up on plinths. Of running, arms spread wide, through the long grass of the savannah at the end of our road, startled egrets taking off in flight. Of being up trees, knees scraped on rough bark as we sat in the crooks of mango branches, tearing into the sweet yellow flesh with our teeth, sticky juice running down forearms.

I must have been constantly dirty, but never felt it. Yet for some reason I cannot understand, this makes me feel soiled, laying out the neat rows of beans, trying to get the spacing right between the plants, helping the children to firm them in without damaging tender stems. When we have finished, I turn to the foxgloves with relief and delight. I notice this split within me, that growing for flowers is good, and clean, while growing for food is dirty, but cannot make sense of it. There is no time to among the chaos of the children anyway. We carefully place the young foxglove plants at the back of the garden beds, scattered in the spaces around the blackcurrant, which is now in full, vibrant bloom. My heart sings to see the vivid red of the currant blossoms, whose leaves I have been drinking in tea most days. In my imagination the foxgloves will tower above it to bring Oli much joy in months to come. The flowers are powerful medicine for an irregularly anxious heart.

***

Sowing seeds and looking after our young plants occupies some of the endless homeschooling time with the children, but when they are playing quietly I increasingly sneak out to the garden on my own. It is intense, and intensely lovely, being with them all the time, but the constant interruptions and questions and observations out loud leave me with no space of my own to think. At times the inability to finish my own train of thought makes me feel as if I am going mad. Stepping out into the peace of being alone in the garden seems to free something in my mind.

In the snatched moments between preparing food, serving and eating food, clearing up food, and now learning about growing food, I wander around the garden thinking about flowers. I want the space to be beautiful, and for me that means trying to have as many things in flower for as much of the year as possible. There are the plants in pots that we brought from our old home, and another group of more recently acquired pots of things that I bought in the last visits to garden centres last month. But before I commit anything to the ground I spend a lot of time looking at the garden, seeing how the light is starting to change with spring, noticing where the ground is wet beside the stream, or where it is unusually dry, from the water's route being constrained in its manmade channels. I am not taking a very planned approach to the aesthetics of the garden, not thinking too much about which colours or leaf shapes go where. I enjoy well-designed gardens, but in this space that idea seems too fussy. I get the feeling that this garden calls for something a bit messier, more loose and free and undone. Or perhaps the current bewildering state of our lives makes me want to plant with wild abandon. Regardless of how it will turn out aesthetically, I still want to consider each plant's home, hoping that I get its particular environment right so that each plant can flourish.

On my mother's birthday I head out to plant the daphne.

The children and I have video-called her, sung happy birthday across screens, and now I need somewhere to put my deep sadness that I cannot celebrate this day with her in the flesh. I take it to be buried in the garden.

The *Garrya elliptica* next to the front gate, whose catkins I had raided for our Christmas table, and which had looked as if it was dying on our arrival here, had finally succumbed. Oli dug it out, and the gaping hole in the gravel next to the front gate had started to bother me whenever I opened the door. On one of my first trips to our nearest garden centre, I had wandered the aisles with a trolley, going round the dispiriting midwinter displays not really sure what I was looking for. And then there it was, *Daphne odora Aureomarginata*, the 'queen of winter-flowering scented shrubs'. Somehow I knew it was what I had been seeking.

I am finally getting round to planting the daphne out. I put some compost in the hole Oli left, and firm down the plant into the gap. Once everything is watered in and tidied up I look at its glossy leaves edged with white, and imagine the sweet flowers that will grace this spot in winters to come. The tag says that it will flower from November to February, and I hope that it will bring me pleasure in those wettest, most dismally English days of my birth month, which I shared with my grandmother. It feels important to place it here, at the entry to our home.

***

Later, while the children and I are out on one of our daily walks along the mossy path that leads through the woods above our house, I am caught by one of my favourite smells. The unmistakable sharp scent cutting through the rich earthy loam of the mouldering leaf-fall beneath the trees. Wild garlic.

A few yards on we find the patch, starting next to the path: broad, strap-like leaves rolling down the bank towards the field below. The wood has been full of ever-green colour throughout winter if you only looked for it – carpets of dark ivy, small rosettes of perennial wild plants half buried by the dead stems of the previous year's growth – but this green is startlingly emerald. The wide, lance-like leaves cascade down the hillside away from us, seeming to light up the dark woods with their glow. I pick a leaf and crush it between my fingers, double-checking its identity even though I am already sure. Wild garlic is one of my favourite spring foods, and was near-impossible to come by in our old home. It feels almost unbelievable to find this generous patch here, many metres long and wide, freely offered for the taking.

Excitedly, the children and I go home and return with a trug. We climb down the bank away from the path edge, and even being careful to take no more than one leaf from every few mature plants, our trug is full when we are still only a fraction of the way through the space.

I find myself unexpectedly emotional. Voice trembling, I make the children pause and thank the plants. I am overwhelmed by their generosity, and feel the first stirring of an understanding of how this land might nurture and nourish me, and provide the strength for me to continue to hold all that I must.

We spend the afternoon making batches of wild garlic pesto. The children and I washing the smooth leaves, piling them into the food processor's bowl and watching them satisfyingly blitz into a bright green paste with pine nuts – walnuts when we run out of those – hard cheese, olive oil and salt. We freeze the excess jars, future-proofing against lean and hungry times to come. We eat some that night, a simple meal with pasta, but I feel the keen emerald green of this wild tonic cleave against my tongue, and begin to wake something in my winter-sluggish, terror-frozen body. And I can begin to fathom how I am really going to make it through the next few days, weeks, months of this life-threatening crisis.

Oli comes home with news. Their work patterns are shifting drastically; he will be on call much of the time, more evenings and weekends away from us. But, para-doxically, this offers us more of his presence in the daytimes, when we have missed him dreadfully. He will be able to share with me some of the homeschooling, some of the day-to-day, mindlessly repetitive work of the house. I will no longer be abandoned at home every day with the children, and our terror. And this will free me just enough from the prison of drudgery, just enough so that I will be able to hold my splitting edges together.

The next morning I wake early. I pull on old jeans, a worn jumper, muddy boots and my gardening gloves. I go to the bottom of the garden and begin to haul sacks of compost up the several flights of steps to the top. I feel stronger already. I begin the process of spreading the richly rotted material around on the too-empty, too-grey, too-dead ornamental beds. This is shit composted into

black gold. I shall need to order more. I feel small seedlings of hope begin to awaken within me, take root as I
caress the soil, blackly blanket the garden's skin.

Judas Tree

IT IS THE most beautiful spring I have ever lived through. Clear blue skies day after day, warm enough to bare limbs outside routinely in April. We eat every meal outdoors, moving from the sunny patch on the lawn behind the kitchen at breakfast, under the shade of the freshly leafed hawthorn on the deck at lunch, to the radiating warmth of the stones on the patio behind the conservatory at dinner. I murmur an offering of thanks for the uplifting weather, for the space to enjoy it with every meal – it feels like a prayer.

Seduced by the caressing warmth, the garden begins to wake up. There still seems to be a disappointing dearth of herbaceous plants in the beautiful, stone-landscaped terraced beds, but the shrubs begin to leaf out, transforming from unrecognisable sticks into tree peonies, wedding cake cornus, weigela. The evergreens that I have disliked all winter above the miniature waterfalls whose music provides a constant soundtrack to the patio outside the conservatory begin to come into flower. With the spring comes their blossom, and their redemption.

The new season's leaves of the shrubby germander turn out to be prettily silver, as if the plant is permanently sugar-frosted, and set off the sweet, dusty lilac of its small, orchid-like flowers perfectly. Next to it, the genista explodes into bright yellow bloom. I have never thought that I much enjoyed yellow flowers, but these sunny, pea-shaped ones, contrasted with the pale, silvery violet tones of its neighbour, give me unexpected pleasure. They are sun and moon intertwined. I pause, and smile, reach out and gently stroke their petals, every time I climb the steps past them.

On the other side of the path, a gangly collection of sticks turns into a lilac. I am giddy to see it, besotted with their scent. As I cut a few stems of their blooms for the house, I could clap my hands and laugh with delight like a child. More than a decade ago, one of the first houses that Oli, my then boyfriend, and I shared was a small country cottage with a wild, barely tended garden. In that cottage's front garden lived a pair of lilacs, arching over a bench. The owners were happy for us to play in the space, and so on sunny weekends we would plant some things, badly prune others, before sitting under the lilacs as the day drew to an end. We returned to that garden after our wedding in Trinidad and sat beneath the lilacs wondering about future children. In that garden, I dabbled in creative writing for the first time since childhood. In that place, the seeds were planted of dreams of a one-day country cottage of our own.

As delighted as I am to see the lilac, and feel the tender, full-circle flood of memories that it brings, it is still not my favourite spring-flowering shrub in the newly awakening garden. The most magnificent lives on the steep bank above the germander and genista, and is one that I have not knowingly seen before.

As Easter nears, tiny, soft pink flowers, almost sweet pea-like, with a deeper purple at their base, cover its leafless branches. It looks like a swarm of butterflies has landed on the small tree, and perch there, delicately fluttering in the breeze, until at some sudden signal they will lift off again, and swoop over into another garden. Swoop across the seas like the spectacle of migrating monarch butterflies to land on the pink poui trees of Trinidad's Northern Range, which also flutter into a tissue pink froth of blossom at this time each year. At home it would be nearing the end of dry season, rather than spring, the red soil even more parched than my grey garden is now. The glory of tree blossom – pink and yellow poui; red flamboyant; orange immortelle – heralds the rain, deciduous trees in that tropical rainforest zone shedding their leaves in response to drought, rather than cold. In this warming garden of mine in England, I look the prettily pink tree up in my plant ID app, and discover it is a Judas tree, *Cercis siliquastrum*. The heart-shaped leaves follow after flowering, also giving it the name love tree. Love and betrayal, how apt for Easter flowers.

***

The magic of Easter has always been my favourite – extremes of death and rebirth, all in one chocolate-coated holiday weekend. I was raised in a deeply Catholic setting. My father, born a Seventh Day Adventist but who gave up that church on leaving home, still retains a profound knowledge of Bible verse. My mother, immersed in Catholicism, her father a lay minister in the church, her mother in the choir. The scandal of my mother's belly swelling while my parents were unmarried students, and their Adventist wedding shortly before my arrival, perhaps forgiven, never forgotten. In true Catholic fashion, I still carry the guilt.

I spent twelve years at a Catholic school, the secondary years girls-only, and was taught by Dominican sisters, one of a few local chapters of nuns. It was a nurturing school where I felt known and loved, but the devotion to the higher spirit that it demanded, with a repression of earthly desires, was not without sacrifice. We were taught to despise the precious homes that were our earthly bodies, to be ashamed of the dirt of our flesh. We crowned Mother Mary with flowers, but were always aware that we were Eve after the Fall, debased, our highest aspiration to cut ourselves loose from this sinful, dirty place we temporarily inhabit. The ultimate conquest of our colonial invaders being the severing of our souls from the daily miracle that is life on this earth.

Christianity was closely linked to slavery. In the early days of the development of the plantations, European indentured labourers worked alongside African slaves, both receiving horrific treatment, but with a much more fluid social position that was not yet set in an ideology of racial inferiority. But to prevent the groups banding together in successful uprisings against their masters, a system of preferring the Europeans over the Africans was instituted by the plantation owners. This was initially based on faith, with Christian lives considered more worthy than non-Christians, who were seen as subhuman. Over time, and influenced by ideas such as the writing of Edward Long, a slaver who published the *History of Jamaica*, the classic text of pseudo-scientific racism, this faith-based hierarchy became a race-based one, with a shift in the privileged class from being called Christians to Whites.

In my childhood I was a profound believer, thumbing the rosary in my pocket, Hail Marys never far from my tongue. I feared God and liked Jesus, who seemed a kind man with sensible teachings, but Mary was the one I truly worshipped, a slightly blasphemous bent to the primacy of her place in my affections. It was Mother Mary that I turned to in my childish distress. It was she that I chastised when I turned sixteen and learned the church's views on contraception and abortion and the strict policing of sexuality. She that I berated when I lost faith in the institution run by these holy men that seemed hell bent on suppressing my womanhood. But it was too late for my early childhood ease with prostrating myself in the dirt. I had repressed those urges. I had risen above such things.

For years I floundered in what I called atheism, but was really scientism, the redirection of my devotion towards those White men in white coats who would teach me the truth of the world. I believed in my studies fanatically, until I was working as a doctor with real patients and realised that, no matter how much I applied the Word of Medicine, there was still so much I did not know, that we could not know. So many mysteries of suffering of body and mind that seemed unanswerable when studied under the limited light of these truths. And the default professional response to the discomfort roused by inevitable uncertainty seemed to be a disdainful dismissal of any concern that could not be contained within a neat diagnostic box, cured with the clean cut of a knife or a pristine pill.

My own prostration as a patient was my biggest test of scientific faith. No investigations or examinations could answer the mystery of why my husband and my seemingly healthy young bodies were infertile. My acupuncturist, practising the techniques of a different faith, offered a clue, and deep within my body told me the answer, but I was afraid to hear it. Unwilling to make the sacrifices at the altar of my career that were called for.

One Easter I stood in the church in my in-laws' village, attending the service we had been invited to as part of the family's weekend celebrations. We had been trying for a baby for five years with no success. It had been many years since I had even had a miscarriage; my body now gave no sign that it might be capable of carrying a baby. When called in the service to a time of prayer, I bowed my head and thought of the goddess Ēostre.

I had begun to be curious about the Indigenous religious festivals of this island, of the ways of worship that pre-dated Christianity. I knew very little about what rituals and ceremonies there might have been on the islands of my childhood, but imagined that they would have been rooted in that very different landscape from which I was now cut off. So I turned to the old rituals of this island on which I was trying to build my home.

I was intrigued by the Celtic wheel of the year, and the way it seemed to punctuate even the darkest times with regular points of light. It was an ancient calendar

that marked eight annual points with seasonal celebrations. I had a growing sense that leaning into the seasons,
embracing even the cold and dark ones that I dreaded,
might help my sense of belonging and connection to this
place. I started venturing out into the landscape every
six weeks with Oli and the children, seeking in some
informal way to mark that point in the year. We took
small picnics, gathered flowers or leaves, fruit or twigs,
and used them to make something beautiful. I was taken
by the stories of the goddesses, intimately intertwined
with the elements and the plants, powerful mythical
figures deserving of worship in their own right, unlike
the subservient image of Mother Mary.

Ēostre, spring goddess of the festival Ostara, held at the
vernal equinox. A time of the renewal and re-emergence
of life after the death of winter, a celebration of fertility.
I held a wordless plea in the silence of my mind, standing
in a Christian church but directing my unformed thought,
my bodily desire, to much more ancient forces. A prayer
that my womb might hold the baby to come from the IVF
cycle we were due to start at my next period. A hope for
my fecundity.

There is debate as to whether Ēostre was a truly ancient
pagan deity or simply invented in the ninth century by
Bede. But the two lines on the strip of plastic that lay
before my disbelieving eyes a couple of weeks later
restored my bodily faith. My womb was a home to that
baby; the magic of Easter gave birth to my son.

***

One day I notice that the beech hedge just budding into this year's fresh growth has created a second autumn, and filled the little stream in our garden with last year's leaves. I set about clearing them out, worried about one of the channels that holds the stream at the places it disappears under the path becoming blocked. A few days later, my fear is realised – a new waterfall has appeared on the steps that lead up from the patio. It cascades down them, turning the steps to ice rinks in the morning frosts. As it still has not rained, we worry that this excess of water means a broken underground pipe, or perhaps a burst water main. Still at home, the initial three-week lockdown unsurprisingly extended to six, we wonder what we can do. Then I remember a small miracle.

Over the Christmas period, when my in-laws were all round, my mother-in-law came to find me in the kitchen. 'There's a man outside the gate who says he built this house.' Her pursed lips conveyed her scepticism.

We went outside to see the purveyor of this bold claim, and met Jim, who turned out to be a stonemason; he had been employed by the previous owners to build the garden's steps, patios and terraces over the course of many years. He told us that he was hired from time to time on day rates, and how, bit by bit, he helped carve this garden from the unyielding clay. He showed us the details he was proud of, was sure to point out parts where others stopped paying him and took over that were not finished to his liking. He talked about the garden as if it was a favourite child. He left his number when he went. I jotted it down, waved him off then tucked the piece of paper into a kitchen basket, and our encounter into memory alongside the growing collection of stories of country life.

I go to the basket now and find his number. Some essential services have resumed functioning. We are worried about the size of the leak, the impact on the integrity of the steps, and the implications if it involves the water main. I give him a call. Would he be able to come, lift the stones over the new cascade and investigate the leak? It turns out that he is isolated here, trapped away from the rest of his family, who seem to be overseas. He is only too happy to have a reason to get out – he will come, take a look and see what he can do.

In the process of having the stream's channels repaired, we learn that it is holy. Apparently sprung from the same source as the medieval well on our neighbour's land, to which pilgrims would come in centuries past to heal their hurts. We live near the city of Sulis, named for the Roman life-giving mother goddess of water and healing, and no doubt water goddesses of the societies who lived here before the Romans arrived, but I am delighted to hear of the mystical powers attributed to this humble stream. Although a part of me is not surprised: every day I walk through the garden it strikes me as magical to see the clear water rising unbidden from underground. We had been left a history of the village by the house's previous owners, and I read it now specifically looking for the significance of the water. I learn that the nearby well was pagan in origin, later appropriated by the church and named for the saint who came to be associated with this region. We are even told that the source upstream was once blessed by a pope, though that seems rather unlikely.

But I hear a new chord in the water's constant song now, a note of benediction.

The source of the leak turns out to be a mundane one. The pipe that channels the stream's water under the steps was not broken, but blocked by an old baby bowl. The previous owner's children were teenagers when they moved out, though they had been born here. That bowl would have fallen into the stream and been buried under the steps more than a dozen years earlier, only to rise again when I troubled the waters, stirring submerged memories of childhoods long past in my blundering attempts to ground myself for the ones at present. That it has lain hidden for so long an ordinary sort of miracle.

***

I am in the garden every moment I get. Whenever my husband's shifts allow him to be around to feed the children, log my son in to his class video call, and find ever more creative ways of engaging with their learning, he takes over the role of caring for them and I head outside. Being able to share the burdens of home is freeing. The more time I am able to devote to the garden, the more that time feels necessary to my devotion to my children. A couple of hours on my own tending the space transforms something within me, and I am able to transcend my anxiety and get through these intense days of parenting in a more calm and contained way.

As I am learning how to tend this new space – most days just stepping outside and wandering around until some task seems right, now that I am nearly out of mulch

– the garden also seems to be learning how to tend to me. I have thrown the gardening books aside now. I am learning from the space itself. Someone looking at me would think I spend most of my time doing nothing at all in the garden; unlike most other parts of my life, somehow it feels right to let myself simply *be* in the space. It is uncanny, but I begin to feel like I am forming a relationship with the garden, and in all the time I spend seemingly idly staring at it I am learning something of its language, and letting it tell me what to do. It all feels very uncertain, and most of the time I feel as if I have no idea what I am doing. But something will eventually guide my hands.

Today, I set to clearing a bed right outside the conservatory. Unlike the empty beds on the terrace above the patio, this raised bed built of stone has been left an unruly mess of dead foliage. Something with arching, woody stems has taken over almost half of the space, and something else which has decayed into a pile of large black leaves over sticks seems to occupy much of the rest. Collapsed over all of it is the dried and tangled remnants of some kind of climber. I had bought myself a new gardening tool for my birthday with clearing and cutting back in mind. It is a Japanese tool, a sort of small sickle. I am not sure if something so small will make much of a dent in this daunting wall of growth. But armed with this and some secateurs, I gown up in a robust old jumper, and put on my sturdiest gloves, and climb up onto the bed. I feel like I am about to carry out surgery on the garden, a procedure for which I do not feel remotely qualified.

I start slicing through the piles and tangles of last

summer's growth, and soon settle into a bodily rhythm that frees my mind. I squat, grasp handfuls of decaying plant material in my left hand, cut through them with my right. When I have gathered an armful, I bundle it up against my chest and take the pile over to the garden waste bin. The rhythm of squat, cut, gather feels soothing, and the act of clearing through this tangled mess feels cathartic. By the time I have cleared my way through the whole space, this patch of the garden feels like it can breathe more freely. It seems to sigh with relief. I feel euphoric.

With the old, dying back growth removed, three woody stumps are left at the back of the bed. The long stems of the climber and the piles of black leaves and sticks have come away easily at ground level, leaving behind whatever they were, to emerge later this spring. I feel lighter with the dense black hole of dead foliage gone. There is a shell of a small shrub left near the front of the bed, some wispy brown sticks, which I leave because I like their colour and branching shape. And at the back of the bed has more prominently emerged a tall, spiky plant, with long, sword-like leaves. They look like they should be green striped with yellow, but despite being the only evergreen plant in the bed, the whole thing looks a bit sickly, as well as unexpectedly tropical in feel. I have a cup of tea with my husband – mint that has appeared spreading down the herb-lined steps – and he tells me it is a phormium, or New Zealand flax, that is probably suffering in this cold, shaded garden. We decide that it should probably come out at some point, when we have the time to remove it and a better idea of what should replace it. Meanwhile, once the pot of tea is finished, I cut away most of the dead, brown, curled leaves. They are tough and stringy, full of strong fibres that feel like they ought to be woven into twine if I knew how. Then I empty the last two bags of compost in the bed, and spread it around to mulch the exposed earth. When I stand back and look at the cleared space, I realise I might be projecting my own, uplifted mood onto the garden, but somehow it feels as though the phormium looks happier.

***

A bulk order of peat-free compost arrives from somewhere in the Lake District, one of the few sources still readily available. It looks enormous to me on its pallet, and I set about filling buckets at a time and taking them around the garden. I kneel in worship before the awakening shrubs, blanketing their feet with my offering of mulch. I am slowly feeding the garden, and in that act the garden is feeding me.

It is intimate, this act of crawling around the garden's beds, and caressing her skin with my hands. It offers me a view into the hidden parts of her from new angles. Crouching beneath the still-bare branches of the beech hedge that runs along the road one morning, I notice thin strands rising from the soil. They remind me of tiny snakes as their eyeless heads begin their twirling journey upwards, charmed from the basket of the earth by the warmth of the sun. There is a strange familiarity to them – they make me think of a vine from home, a relative of the morning glory that has escaped into the wild and smothers entire trees with its embrace as it climbs into the canopy greedily reaching for the light. I search around for a while and soon find what I am looking for: the heart-shaped leaves of a vine that has been growing up the middle of the hedge unnoticed among the brown, dried tangles of last year's growth. It is a relative of the same bindweed, purple-green winding tips rising from the fleshy white slumber of underground winter roots, resurrecting everywhere, come to poison my Eden.

I panic to see them. Everything I read and remember about the plant tells me how it will smother my other plants, how impossible it is to eliminate, yet how vital that we do. Recommendations go as far as digging out everything that currently lives in the affected beds and then carefully sifting out every last piece of root, before replacing the topsoil and replanting, or blanketing the soil with weedkiller. I look at the mature hedge, at the tall lemon-scented conifer and other shrubs in the beds, and know that I will not be uprooting the garden as a whole. Nor could I ever poison this precious space. As vitally linked as I am beginning to feel with it, it would be like poisoning myself. So I will be weeding them out by hand. Trying to keep all these ropes of bindweed from rising into view will be a thankless and all-consuming task. But I must do the right thing in eradicating this garden evil, and I set to it with fervour.

I lie under the hedge day after day, teasing out the thin, white roots from the thick leaf litter. I soon learn that I must be patient with my efforts, and hold only a slight tension as I pull on the roots, following along their path with my hori knife. If I get impatient and tug too harshly, the roots snap. I do my best to follow them along to where they dive under the hard, woody roots of hedge, and with a surge of at first frustration, then resignation, and eventually amused respect for the plant's tenacity, I inevitably break them. I discover that new growth will re-emerge manyfold from the ends I leave behind, but eventually they will weaken. I walk in the field just behind our garden and see bindweed flourishing among the grass. I realise that my mission to eliminate it from my garden is utterly pointless, but I persist anyway. My initial panic has faded; there is something meditative and calming about the task. I am wholly immersed – in my body, contorted under the hedges or bent over the old veg beds, holding just the right tension in my left hand as it grips the white rope of root, my right teasing soil away so that I may continue to follow its path, keeping it whole as long as possible. And simultaneously immersed in mind, attention completely absorbed by the task, racing thoughts still and settle. And in the background the rich mesh of hyphae and roots holding together the soil on which I lie, beginning to knit together the wounded edges of myself into one growing whole.

One day I find myself squatting with my bucket of mulch beneath the Judas tree in its full glory. Perched on the slope of the garden rising above the house, entirely covered in delicate blossom, it again reminds me of the pink poui trees that dotted the hills rising above my childhood Cascade home, in flower at this same time of year. I sit in the dirt beneath it. I have not cleared this space much yet: it is a mix of gone-over primroses, some kind of shade-loving, self-seeded sedge, and a spreading mat of dead nettle that I suspect will be hard work to remove when I get round to more intentionally planting this space. For now, I'm just offering the small tree some nourishment in return for its astonishing blooms. I pause a moment to relish them, and to feel all the senses of home that live under its canopy.

Leaning against the trunk, a halo of pink flowers surrounding my head, I realise I am hidden from view from the house behind another unpruned evergreen. Sudden tears well up in this unexpected private moment. I find myself crying into the undergrowth of the garden a lot these days. I cry for all the things that I have lost – certainty, security, safety. I cry out of frustration with the world. Oli has been returning from the hospital distraught by the government's edict to discharge patients to unprotected care homes without testing for Covid-19, knowing that what will inevitably happen is akin to mass murder, but helpless to stop it. I hold him at the end of his shifts and feel his spirit buckle under the weight of moral injury he is having to endure. My stomach churns as I see an unnatural hierarchy of the worth of human life being imposed, just as it was in the colonies under slavery; the justification of the inhumane sacrifice of one type of body over another. In my Black body I shiver, the scars of Imperial callousness to the lives it has trampled in upholding itself tingle under my skin. But it seems that so many people are too deep in denial to see that while they may be a body or two deep from the edge now, without working to value all of our bodies, and keep them safe, they too will eventually be pushed over into the chasm.

And I cry with fear for myself. Reports have begun emerging of the deaths of UK healthcare workers exposed to the higher viral loads of the very sick patients who are hospitalised. My fear of losing my husband to this virus, and the vision of being left to manage this unwieldy house and garden and look after grieving children on my own, cut off from other support by the very virus that might cut him down, overwhelms me. The data highlights the shadow of our government's great betrayal: the exposure of those who are working to save us to the highest viral loads, armed with the lowest protection. Our leaders one great Judas, sacrificing those who would lovingly save us with a murderous kiss.

***

The garden is my safety valve, where darkly entangled feelings that I cannot even always name can be poured into the soil, which seems able to absorb them all. Without it, I bubble over and quickly spiral into patterns of shouting and withdrawal, overcompensation and collapse. It is obvious how much the garden is holding our family unit together, so prioritising my time in it becomes paramount. This lesson of filling my own cup so that I can pour from it to my children yet another that the garden seems to be teaching me.

My favourite time is at first light. I was never a morning person before having children, but watching the first of the day's beams travel over the garden, kissing every leaf as they go, is a source of pure joy.

Light has entered the garden now. It draws me with it. And as the sun sparkles along the stream, the small waterfall into the ponds on the patio a refracted prism framed by moss, it highlights how profound the darkness of the previous months has been. I realise how the garden sits in almost complete shade all winter; apart from a couple of the highest points, the sun never rises above the tree- and roof-lines around us to touch the space. Then at some point around the equinox it appears, and with it the garden is a completely different place. The stone terracing is suddenly warm and inviting, instead of cold and damp, the mossy stamp of 'lawn' behind the kitchen dries out and actual grass begins to appear. Spaces that I thought were deeply shaded are now haloed in bright, full sun. I realise that I must throw out all my assumptions about the space that were half formed in the night of winter, and re-examine it in the dawn of spring.

The clear blue-skied light casts stark shadows. It is only in the light of day that I can see clearly the darkness I have left behind. This is often the way of things, and sometimes I have been unable to see without someone else to shine the light into dark spaces, a therapist to hold my hand as I pick my way through my mind's shadows.

I left my own therapy when I left Oxford, and standing in the warming, light-filled spring garden, I realise how much I miss it. For years I lay on my therapist's couch twice a week, with breaks only for holidays, and brief ones when I finally had my babies. Otherwise, there I was, in her basement womb of a therapy room, examining the shadow parts of myself before emerging, blinking into the light. I understood my depressive tendencies better, my manic defences against them. I grew to see, and to name, my core deep fear of abandonment, my self-sabotaging tendency to abandon others first in order to save myself the expected

pain. I began to connect the creation of those abandonments to Imperial uprootings, where forcibly separating
families, carving grotesquely humiliating rifts between
lovers, and separating us from one another and our very
selves over centuries, had created the most cruel but
effective forms of control. Defence mechanisms of the mind
compelling the repetition of these scarred and branded
patterns even once we were supposedly freed, through
generations to come.

Held in therapy, I could see and name the abandonments
that had directly shaped me: my mother's leaving when
I was eight months old, to complete the master's degree
interrupted by my unplanned arrival; the reluctance of
my father's return – four years of my life before he came
at her ultimatum to the small island he thought he had
escaped. But it was my mother's leaving, in that time
before language, memory and understanding, that seemed
to have – entirely unintentionally – most wounded me.
My mother finished her diploma in Jamaica, and returned
within a year, and was devoted to me, but it was not
enough to claim the title Mama. As I gained the power
of speech, this lifelong privilege was bestowed upon my
grandmother, who cared for me in my mother's absence,
who grew me alongside the roses in her garden. And then
I reached adulthood, and I left them both.

It was only after having my own babies that I could
begin to sense the true depths of these unconscious and
unwitting retaliatory woundings that I had inflicted on
my mother for her abandonment of me. She had no choice
– I knew that. She had done it out of love, making what
she thought was the best decision for both our futures,

choosing the right path to success in her post-colonial time. I was so very proud of her achievements, of the woman she was, and the place in the world that she modelled for me. She was so proud of my achievements, encouraged my ambition to attend university overseas, supported my leaving, and then cried every minute of her return flight home. And so it was that I became very certain that the cycle of abandonment needed to end with me.

It was the long, slow, at first deeply shameful and painful process of excavating the repressed meaning of these truths about myself, of slowly breathing life into the dead, dark corners of my mind, that started me on this process of seeking home. It was only after I had brought into the light of consciousness my simultaneous longing for and fear of belonging, the desire to be loved and fear of losing love, that I could approach the vulnerable issue of homing myself. It was only after I had lain on a sofa week after aching week and felt like a fragile outline of myself, formed of nothing as significant as dust, that could be destroyed by a too-sudden breath, and gradually, atom by molecule, rebuilt myself, forming solid substance from shadow. After I had looked into the sea of churning, nameless dread in my core, and fished the monstrous feelings out of it, one by one, finding them to be not monsters at all once in the light, but the ordinary human pains of loss, and grief, and fear, and envy, and hurt, and love. It was only after all that, that I could awaken to looking at the precious heart of myself, and see it as deserving of belonging to a home.

***

I have never known a spring so warm in all my time here. Nor so dry. Days, then weeks go by, and still no rain. I take to leaving the children's teepee full of toys out on the small lawn behind the house overnight, safe in the knowledge that it will not be rained on – unheard of in my time in England. Beneath the black layer of compost that I am slowly spreading across the clay beds, cracks appear. Fractured fault lines of thirst marking the garden's skin. I assumed that because there is a stream here, because it was all so sodden, and mossy, and damp all winter, that drought would not be a problem in this garden. I begin to learn that things are not quite so black and white.

In the unusually warm spring it is hot, dusty work in the garden. The mended stream bubbles and babbles so invitingly that one particularly warm day I strip off my boots and socks, roll up my trousers as far as they will go and lower my feet into one of the small pools formed out of rock. The water is bitingly cold, but refreshing. I find myself wishing that I could strip off entirely and immerse myself in the water, but the stream is far too tiny here to allow that. Instead I wash as much of myself as I can, cupping the water between my hands. Looking down into the silted-up leaves at the bottom, I can see that the stream is teeming with tiny freshwater shrimp. This reassures me that it must be clean, despite the risk of run-off from the fields that lie above us. This spring must well up from the groundwater far below. I will not risk drinking it, but as I splash the cool, clear liquid on my arms, face and neck, it slakes a thirst.

I have always been a water baby. Born in the watery season of Pisces, swimming was always the sport I most enjoyed. The seas were my natural home, with the constant tumbling waves and fickle rip currents to read, and be wary of. But the most peace came from swimming in the rivers. Most recreational swimming in the islands happens in the form of socialising in the shallows, so a swim in the deeper waters upriver meant gliding through still, dark water, tropical rainforest descending to meet either bank, no souls in sight other than the occasional bird in the canopy, or the eerie prickle on the back of the neck that brings to mind Mama D'Leau.

Before the slavers brought Christianity and its justifications of enslaving the souls of the savages, the island of my childhood had many goddesses. A sister to Sulis, Mama D'Leau was the mother goddess of the waters, healing and protecting those who came with good intentions to her shores. Of course, diving into deep waters carries risks, and an interaction with the Mother of the Water carried the risk of meeting the Styx, if your intentions were not pure. I swam in the Marianne and Grande Riviere as an innocent child and felt safe in these remote, northernmost rivers. I dived off rafts that local children poled upstream and floated, eyes closed, in the soft warm of the river mother's watery womb. I always emerged feeling born again.

In my English garden, I wash myself in the waters of the garden stream as in a baptism. But I make no sign of the cross here. Father, Son, Holy Ghost, these burdens of men will not be my cross to bear. Instead I encircle myself with the water, crown myself with flowers from the plants around me. I find myself offering a sort of prayer for healing and protection to whichever minor spirit might watch over this small stream that will somewhere meet the sea. Insurmountably vast acres of ocean separate me from my mother in this moment. Tears well and fall, and they swirl with the stream to the same sea that will meet the waters of my mother's home, watched over by Mama D'Leau.

I look into the pool, and my distorted reflection looks back at me, crowned by a drift of fallen Judas tree flowers. She looks like the statues of Mother Mary that in my childhood we decorated with flower crowns. She looks like my mother.

Wisteria

A HEAVY SCENT drips through the open windows of the lounge, accompanied by the buzzing of bees. The pale, twisted branches that spiral around the metal support and up the front of the house have revealed themselves to be wisteria. There appear to be two plants, each rooted at either end of the house. I cannot tell which might be older from looking at them, but judging by what I know of the history of when the house was originally built, and then extended, the one which has burst into bloom and scents our living room so heavily must be the elder. The thick trunk that climbs onto the newer part of the house is in dense leaf, but no flower buds have appeared. It reaches out to its neighbour, and their leaves overlap and embrace each other. I inhale the rich scent with half-closed eyes and revel in the wonder of this, that I now live in a wisteria-covered cottage, the idyllic country-side dream.

The long racemes of heavily scented, pale lilac flowers remind me of something, but I do not place it until I am sitting in the window looking out at the blossom while texting with my mother one day, and suddenly a vivid mental image comes to mind. I am standing in a long, straight avenue of vining plants pruned and shaped to standard lollipop form, and they are in abundant bloom, covered in racemes of sweet-smelling, pale purple flowers. It is May in the petrea avenue at the Royal Botanic Gardens, and the queen's wreath, or tropical wisteria as they are also known, are in full flower. They bloom for May, for Mother's Day in Trinidad, for the month of Mary in my old school calendar of religious celebrations. We have already had Mothering Sunday here in the UK a couple months ago, but here I am texting my mother to wish her a happy Mother's Day back home where the *Petrea volubilis* will be in flower, and here is the wisteria wreathing my home in bloom for the occasion. It comforts me to see it.

The connections between my Caribbean home and this one have seemed profoundly strange, and unexpected, but as I communicate with my mother, thinking of these maternal plants, the links between them begin to fall into place in my mind. England was modern Trinidad's colonial motherland; the Royal Botanic Gardens where the petrea avenue grows was a colonial creation, established in 1818, as was the introduction of wisteria to this country, brought in 1816 via the Inspector of Tea for the East India Company, who was acting on commission by renowned botanist (or botanic thief) Joseph Banks. The flowers travelled along Imperial trade routes, which moved plants and people around the world, so often violently against their will, like poisoned umbilical cords that bind places together to this day. And yet, despite the ambivalence that may grow as we come to know them as adults, their faults as well as their care deeply embedded in our flesh, so often we still love our mothers. Even hate not love's opposite, but its perversion.

***

The weather is still glorious, and we do everything in the garden. We are nearly two months into an ever-extending lockdown, and the sense of surreality has not abated as time has passed. The days are unfailingly hot, sunny and dry, like we have been transported from damp and dismal England to a foreign country through the unusual act of locking ourselves within our houses. The isolation is intense, and in the almost total absence of interaction with others I am being increasingly forced to confront myself.

With the schools still closed, I am mothering more intensely than ever before. As well as mother, I am cook, cleaner, gardener, teacher, teaching assistant. I am playmate, worried wife, concerned daughter, extremely distant friend. But mostly I am home. I am Mama, Mama, Mama, a thousand, a million times a day, the children at their most content when they play together what feels like right beneath my feet. In the kitchen of our life I dance, balancing a hundred boiling pans and sharp knives, doing everything I can not to trip over my little ones and stumble, raining all the danger that I precariously hold at bay upon them.

Everything feels freer in the garden, and so we spend nearly all our time there. Apart from sleeping and cooking, we abandon the house and do most of our living outdoors. This is especially because the heating engineers have returned, as the provision of essential services are slowly resumed. Armed with more sophisticated PPE than my husband has access to, they resume the task of connecting the house to the pipes drilled underground, and after wiping any surface that we might have touched overnight, we retreat outdoors and stay well out of their way.

Our feet are grass-stained, and we all have mud ingrained beneath our nails. In the morning, I frequently brush leaves and petals from all our pillows, as we shed the bits that plants have left in our curls the day before. Our skin grows browner, and the children hold their arms against mine, comparing our tones. My son the fairest, my daughter now as brown as I am in the winter months, mine the darkest. All of us glowing golden from the sun's kisses. Should we stop moving, and lie still on the soil for long enough, I feel our feet would put down roots. We are growing into the garden.

We are living through a mass trauma, but the garden seems to have turned on something in my DNA, reactivated some deep remembering of relationship with the land. And I have found that, with a joy as deep as the pain of the moment we are living through, the land remembers me. Every morning as I step out into the garden I feel overwhelmed with gratitude for its all-encompassing grounding, to which I give myself over, to hold body and mind safe. The garden holds me on its lap, cradles me against its dark skin, gently brushes fingers of leaves and petals and twigs against my cheek and through my hair. Sometimes it feels as if the garden loves me.

**

The soil is warm, and it is hard to imagine a damaging frost on these sunny days, so I turn my attention to filling up the mulched but empty beds with plants. The terraced landscaping of the garden, and the way that so little of it is given over to lawn, means there is a lot of ground to fill. I spend hours trawling websites online, attracted to the smaller, specialist nurseries, who are still finding ways to deliver their plants. I place one order, then another, online shopping baskets filled up with somewhat random selections of plants in nine-inch pots. I am still getting to know the garden, or even what I am doing, so I am experimenting. I buy things that I like the look or sound of, that from their descriptions might perhaps work in what I am beginning to understand of our garden. I buy things on instinct, and hope.

Eventually, orders arrive. I unwrap boxes packed with small plants wrapped in layers of newspaper with all the excitement of a child opening presents at Christmas. Their names are strange to me; I am learning a new language. Orange *Epimedium x warleyense* because orange is one of my favourite colours, scented *Maianthemum racemosum* for the shade, *Filipendula rubra 'Venusta'*, pink and fluffy and also scented, for a sunny spot by the stream. *Pimpinella major 'Rosea'* for I am not sure where, because the name amuses me; bright blue *Corydalis flexuosa* because it is a spring ephemeral, and I do not know what that is exactly but it sounds magical.

I take my time about planting them. I have read up about the plants and applied some logic in ordering them, but once they arrive I find myself being guided by something else in deciding where to plant them. I wander round the garden with the pots, placing them here and there, shifting them a few inches, until eventually I pull out my knife and bury their roots where it seems to feel right. I am creating a garden based on my instincts.

The whole process is joyful. In delight I place them around the borders, bed them in, blanket them with more mulch, water them with my dreams. Each plant is a present joy, and a future hope. It is compulsive, and soon I have filled this first space that I am tackling with plants, ones that I have bought and, when they seem big enough, ones that the children and I have been nursing along from seed. There are dozens of seedlings, and I cram them into the beds, led by instinct and whim and fancy. I place edibles among the ornamentals, in what I would like to believe is true potager style, every new rooting a much-needed dose of hope and joy and delight.

***

The house is porous. It has unusually large windows for what must have originally been a humble worker's cottage. In winter, ladybirds came in through small gaps and colonised the windowsills; now the tendrils of the wisteria creep in and along the walls, clothing the house inside and out. There are doors on every side that lie open in the ongoing good weather, and, as in many old buildings, a series of irregularities through which the outdoors constantly makes its way in. The biggest at the moment is a hole in the cellar wall through which the now-removed boiler flue released its gases outside. We temporarily blocked it up with some paper, and one day Oli goes to inspect it with an eye to filling it in more permanently, then comes to find me, excitedly whispering 'Come and see.' I follow him round to the side of the house and, at his urging, stand some way back, looking at the small gap, waiting. Soon, a robin emerges, flits away, shortly returns. I smile as I understand and, when next the robin leaves, carefully peer inside the space to see for myself. A nest has been built inside the gap in the stone, and is occupied by a handful of tiny, hungry, chirping chicks. We move away quickly, so as not to startle the parents into abandoning their nest. The hole they have nested in creates a cold draught that sweeps up both flights of stairs in the house, but my heart is warm at the realisation that our family is not the only one calling this building a home.

One evening, while we are sat reading the children a story before bed, we notice something plunge from the roof outside their window. One, then another, until we realise that a colony of bats seems to be coming and going from just under the roof. We all pile outside and crane our necks to peer at the roof-line. Oli notices a tiny gap just behind the gutter above the children's bedroom window – the bats must use this space to get into the loft. Later that week, one gets into the children's bedroom when they complain about the warmth and I open their windows a crack. There is much shrieking and laughter until Oli manages to gather the no doubt terrified tiny creature up in a blanket and release it outside. Later that night, we read up about bats. Judging by their size, they must be pipistrelles, who have come for the summer to have their babies above our heads. We read that they should disappear in winter and may never return again, but every evening, when our babies go to bed, and the bats emerge to hunt so they can feed their offspring, we watch their swooping, diving flight.

This house, so permeable with the landscape it sits in. Indoors and outdoors blurred, so that the garden even grows within. Our house holding us all safely within its embrace, a home to so many mothers.

***

I had planned to be a relaxed and pragmatic mother. A calm and organised one, who was lovingly present for her children, but logically on top of things. I knew that I would enjoy spending time with the children, guiding them in crafting, or baking in the warm glow of our imaginary sunlit kitchen table. I was sure that I would adore looking after an infant, breastfeeding them every few hours and changing their washable nappies, before putting them to lie down on their stimulating play mat while I got on with a few chores, or maybe had a nap of my own while they snoozed in the Moses basket. I knew that motherhood would be hard work, but was somehow still entirely delusional in my imaginings of it. I realised later that the image of motherhood I had held in mind was that of the Virgin Mary, placid and gleaming, a shining cherub at arm's length tenderly balanced on her knee.

Motherhood savaged me.

When the time came for my son's birth, miraculously he was born in the way I had imagined him coming into the world. I squatted and panted and moaned my way through the most intense physical labour I had ever endured to a triumphant climax. He was born into a body of water. I had spent my entire pregnancy with him seeking immersion. After many years of not finding the time or inclination to swim very much, this pregnancy brought a resurfacing of my childhood love for water. In its weightlessness was where my ever-changing body felt most at home, the rapidly shifting boundaries of myself finding least resistance in its fluid embrace. And then, about thirty-six hours after I had ridden the wave of the first contraction while cooking a Trinidadian lamb curry that I had had a sudden urge to make that day, I squatted in the pool of water deep in the hospital's belly. I was held by my husband and a close friend from medical school who was acting as my doula – not at 'home', where we had started this labour and I had hoped to end it, but at the hospital that had housed our training, our medical home.

I had been set on a home birth since before becoming pregnant. Obstetric friends who were used to seeing the births where it all went wrong thought I was crazy seeking to do something they saw as so dangerously primitive. But I remembered my stint with the midwives in medical school, assisting at the medically straightforward births, where this ordinary, everyday act that kept our species alive felt every time like a miracle. I remembered the shift in feel of the room as the unspoken signs of the baby's imminent arrival became clear, the way experienced midwives would discreetly begin to prepare, quickly but calmly. And I remembered the tremendous change in the room's energy as new life made its arrival. A charge of invisible light that suffused all of us who were privileged enough to witness it, that briefly lent itself to us so that we were lit, shining from within. A surge that momentarily tilted the universe on its axis and felt powerful enough to change everything. And then an overwhelming flood of joy as through the portal of a mother's body the number of occupants in the room was increased, and a baby met their mother. I was moved to tears every time, the foolish medical student trying not to cry over whatever task I had been given, overcome by the radiant possibility of life.

I wanted that feeling to mark my arrival in motherhood. My rational research supported the idea that the best way to safeguard a medically straightforward birth for myself was to plan for that birth to happen at home. The urge to birth in a cosy, homely corner also seemed to come from somewhere beyond logic. The idea comforted me, and made me feel closer to our little house in Oxford. This place we had lived in, but in which I had never felt grounded.

My husband was not comforted by the idea of having our baby there. He wanted to be held in the arms of our mother hospital, where we had spent so many years of our training, and on the maternity ward of which he had done his placement. That familiarity soothed him, and ultimately provided the container he needed to hold me. Nervous while we were at home, his ease was apparent when we eventually landed in hospital even though my labour was progressing, from a medical point of view, fairly uneventfully. For me, it was one of the defining events of my life. The midwife who had been sent out to our home was inexperienced; it was her first unsupervised home birth. Faced by three doctors – myself, my tense husband, and my best friend – she sent us to the safest place she could think of; the place that had birthed my identity as a doctor. And so, in a way, my son was born at home.

Squatting, belly submerged, attempting to peel my hips as wide open as they would go, guttural noises that I did not know I was capable of issued from someone who seemed to be me. As I felt my pelvis crack open, I was overwhelmed by the surreal sensations of the head and body of another human descending through me, to begin a new life outside. As I felt the child who was to be my son squirming alongside my own belly's contractions, something in my mind cracked and split open as well. The delusions fell apart as a primal roar and a baby's head bulged from the very core of me. I came back home to my animal self.

As a very young child I had crawled into our dog's kennel, the only human trusted to be present as she gave birth to a litter of pups. I must have been about four at the time. She was a German shepherd, as big as I was, and much stronger. And yet she let me sit with her still-wet pups, and watch as she ate her placenta, and then the small, deformed body of the runt of the litter.

Somewhere between that experience, and being split open in the moment of birth, I had forgotten, and had lost my way. I believed the lessons of Christianity, of 'civilisation', though my black body was barely seen as civilised. I believed that I was not really an animal, but a human, some special creature that sat on a clean, white plinth above the rest of nature, and held dominion over it. In the bloody pool of my birth as a mother, as I scooped my son out of the dark water and held him against my chest, crying and laughing with a strange recognition as

I said, 'It's you! Oh, it's you!' over and over again, something of my instinctual, animal nature was reborn.

***

Motherhood returned me to the garden. What is a garden but a womb, a space fecund with potential?

My first child was not an easy baby, colicky and with reflux, relentlessly hungry and a terrible sleeper unless held in arms, but as I grew to know him I began to understand that he was remarkably clear in letting me know his needs, and things felt easier if I followed his instructions.

He hated the car with every fibre of his tense, balled-up, screaming little body. And I was unable to bear not being able to hold and try to comfort him when he was in distress. His screams were excruciating, and intolerable. One ill-fated trip out a couple weeks into his life, still rent and raw from childbirth, I made my husband stop the car suddenly on the side of the road as I nearly threw myself into moving traffic, so desperate was I to get my son out of the car seat and soothe the awful screams that seemed to rip at something within me. I was eventually talked back into the car and, sobbing, managed to tolerate the remaining few minutes of the drive.

Our destination that day was Waterperry Gardens, a country estate, garden centre, horticultural school and café, which would quickly grow to become one of our most visited places. Once there, my son was soothed, and so was I, held in the landscape of green fields with a river running through it, curious cows wandering up to greet us over the fence along wooded paths. The instruction was clear: spend more time outdoors. And so I did, the parks and gardens and woods in which I spent my first maternity leave holding me as I grew into my role as a mother, and as I held my son. When he started walking at nearly ten months, his first act of the day would be to toddle to our front door and hammer on it, begging to be released outside, as I too slowly gathered up the things we needed to venture out to the nearby parks. I wished for a garden that he could safely play in, but in its absence the public green spaces in which we spent most of our time held me like the faraway mother I longed for.

***

I had stopped thinking of plants or gardening when I first moved to England. The student life, the new culture, the demands of my course were all absorbing enough. I floated through a surreal ether of bops, balls and black-tie dinners in between labs and lectures. None of this was a medium in which I could ground myself. Instead, unconsciously I would be drawn to the glasshouses at the Botanic Garden whenever I felt low, or homesick. In the warm humidity, softly familiar, the cold in my bones abated a little, winter-dry skin loosened, and I breathed in the smell of home. I would cocoon myself among the dew-jewelled large leaves at times when my family felt impossibly far away, and offer the salt of my tears to these fellow foreigners, trapped under glass as winter lay oppressive and grey.

The only plant in my care at that time was a lucky

bamboo left in my student room by my mother as she said farewell – the room described in the letter from the college admissions tutor as having a view of a lovely but mysterious tree called a magnolia, and which on arrival opened into the tree so directly that I sometimes felt I lived in a treehouse. I had intimate instruction in the divinity of the magnolia that spring. Through some kind of maternal magic, the bamboo survived six years of neglect and nomadic student living before, in a mother's fate, being unceremoniously abandoned when I moved into hospital accommodation to begin my working life. A life marked by even greater upheaval, moving from new hospital to new job to new abode every few months – no point putting down tentative roots only to wrench them free at the next enforced removal. I found that I already knew how to travel light, the only baggage carried within.

The next time I needed my mother, I bought a plant. Heavy with the needs of my two much-longed-for babies, in those early days of learning how to be a mother the thing I desperately wanted was my own. Instinctively, I looked for her in the reassuring plants of home. For my birthday, when the children were finally old enough for me to feel secure leaving them both with our childminder – the salaried embodiment of the grandmotherly presence I craved – I took myself off to a garden centre. My husband and I had been gradually getting into gardening outdoors, in an accumulating number of pots squeezed onto the tiny patio outside our front door. Apart from a few bulbs at Christmastime, houseplants had never appealed; but on this garden centre visit, for the first time I paid attention to the indoor plants, breathing in the warm, humid air as I wandered between dripping shelves crammed with a selection of familiar, broad-leaved plants. I did not know all their names, but recognised them as kin. Their leaves brushed my skin as I walked the aisles and understood that we had all been uprooted and transplanted to this indifferent spot.

In what I can only attribute to a brief and focused form of madness, I had soon acquired close to a hundred house-plants. I bought books about them, and listened to podcasts for tips on how to care for my new charges, but mostly I looked at them. Early every morning, before anyone else awoke and demanded anything of me for the day, I would pour myself a coffee and, with the hot black liquid on my tongue, wander round them, stroking them, admiring their form, poking my finger into their soil to gauge when next they might need watering. I tended my plants as I wished someone might tend to me.

In my ministrations, I learned that, for all that I adored their lush beauty, I did not have the capacity to nurture anything as fussy as a calathea, but that the plants who could hold their own alongside the needs of my children were the ones who would be my companions. I learned that I enjoy a communicative plant, one who has clear signs of letting me know early on when they are thirsty, or infested, or otherwise unhappy. I learned that the most robust plants under my care were not always the most obvious candidates, like the reputedly fastidious maiden-hair fern, which surprised me with its continued existence years on. And as I knelt with my very small children, tending to one need or another, I would briefly lift out of our still-merged selves and look at my tropical houseplants.

Both my mother and grandmother had enormous philo-dendrons in their living rooms; they both tended their tropical gardens. I longed to be held by them as I cared for my babies, and tended to mine.

***

The hawthorn tree that overlooks the deck has come into flower. Unprompted, my daughter declares it the mother of the garden. I am making ever more garden teas, wandering around with the strainer most evenings looking at weeds, identifying plants on my phone and plucking a daily-changing mix of those that have herbal uses and seem to draw my attention for my teapot. Intrigued by her perception of the tree, I read up about the hawthorn, to see if I can make a pot of tea with the flowers. They are supposed to be good for the heart, and the intensity, anxiety and fear of the last couple months has taxed mine to its limits. Its walls have stretched to hold my fears for Oli at work, fears for our vulnerable parents, and to create an entire world to hold my children within. Every fibre of my heart muscle is tired. I read that hawthorn has traditionally been used to make the heart stronger. I go to fetch my teapot and pluck a few of the flowers and leaves of this mother tree to sit under boiling water. I sip the earthy drink that it makes, and feel my heart swell with gratitude for this garden that seems over and over again to give me exactly what I need.

***

As we near the end of the month, and summer swells into view, the garden is delighting me more and more each day. One flowering plant after another begins to wake, and as the colours sing out I fully feel how it is that I live for flowers.

The chorus of colour in the garden grows louder every day. A firework head of coral and orange with yellow centres explodes down by the stream over tall glaucous stems with stripes of mauve down the middle. *Euphorbia griffithii*. The first daisy opens on a messy bank of lank twigs that drape over the edge of the sleepers behind the conservatory. It is bright purple, almost iridescent in the morning sun, with a deep yellow centre. A clump of sword-like leaves next to the stream turns into the most gloriously blue iris I have ever seen. I try to photograph it, but my phone camera cannot capture with its pixels the pigments that my eyes drink in as if parched. I stare and stare at it, awed by the intensity of its colour, as if I have never seen blue before. I have never seen blue like this. Other irises appear too: the sibirica type that are fond of our clay soil, a yellow that I am less fond of, and a pure, spectacular white. And at the back of the messy corner from which the oil tank was removed, a collection of thin stems covering a rickety trellis declares itself to be a clematis, with huge purple flowers the size of my daughter's face, and as beautiful.

The flowers appearing in my garden are big, and bold, and over-the-top compared to the lacy white froth currently lining the paths through the village where the cow parsley is in exuberant flow. They are sultry in their size and intensity of colour. My parents text me pictures of the plants coming into bloom in their garden at home, the planting in that new garden that they have developed without me coming into maturity after a few years' growth. We should have visited to see it this spring. I squint at the bright colours on my phone, favourite plants that they have re-used from the garden of my childhood: foot-long racemes of yellow *Gmelina philippensis*; a brilliant orange hibiscus; billowing clouds of Antigua heath. And some new plants too: lilies planted in honour of my daughter; something I don't recognise with glossy, lipstick-red flowers that look almost plastic in their perfection; and flaming chaconia, Trinidad's national flower, standing proudly front and centre. I wish I could be standing next to my mother, reaching out to touch the flowers together as she tells me about how she chose them.

Instead I am in my garden on the other side of the Atlantic Ocean, a distance whose vastness I am fully feeling for the first time. The vibrant blooms opening one after the other in accelerating succession under this unusual spring's hot, clear skies are a comfort. Something of the familiar seems to have been waiting for me here in this garden. It eases my ache for home. It makes me feel at home.

The flowers slow me down as I walk among them. There are so many! I wonder about responses to the uncannily sunny weather until my husband reminds me of how much I have fed the plants with mulch. I sensed what the depleted garden needed, and it is thanking me a thousandfold for my efforts. I must stop to inspect each beautiful bloom, raised and open to the sun. I bend down to stroke their petals, touch my lips to them as I take in their scent. I stop and raise my own face to the sun, let my shoulders drop and feel my chest open, heart warmed by its rays. I am one of the flowers I walk among.

I have come to the garden like a child to its mother, and it holds me indulgently as I play within the space. Sometimes destructive, I tear through her like a toddler on a rampage, ripping out without thought weeds that probably nurture entire miniature ecosystems. Sometimes we create something together, as I carefully plant into an area with consideration, trying to imagine what my efforts might look like in five seasons, or ten – the first time I have been able to let myself dream so far into the future in one place. Sometimes the garden merely holds me on her knee, absorbs all the intensity of emotion that living through these unprecedented times stirs, soaking it up like rain. I sit in the garden's embrace, and feel more held than ever before. The garden holds me, and allows me to carry on holding my children.

Every day I go out to the front of the house and stare at the wisteria. Each pale lilac, intricate flower opening in succession down the lengthening racemes is astoundingly beautiful, and the scent is a dream. The flowers are as delicious as they look, being used to make fragrant cordials. But the seeds are poisonous, two of them apparently being toxic enough to kill a child if consumed. This powerful combination of sweetness and danger, life and death, feels resonant of motherhood. I look at the elder and younger plants embracing in the middle of the house, and wonder if one has birthed the other; if the younger is just biding its time until it too will be mature enough to reproduce and come into flower. Their thick, sturdy trunks braced against the stone walls seem almost as if they are holding up the house. The plants hold this building as it holds me and my family, and all the other creatures making a home within it.

One day, near the end of the month, I leave the wisteria and walk round the garden, admiring each bloom in turn. The big, bright, colourful ones immediately have my heart, but I am learning to love the more restrained flowers too. The cascade of pure white, bell-shaped campions that creep along the ground and cascade over the point of emergence of the stream; the dainty nodding heads of veronica; a froth of pale pink geraniums. There is a flower for every mood, for every potential facet of my self.

It has been six incredibly turbulent months of settling into this place, through the strangest of times. But on this glorious day, beneath perfectly blue skies, as the magical riot of May flowers take over in the garden, I allow myself to believe that this place really was meant for me. It was the right thing for us to do, to land and put down roots here. I had been unsure of the urge to root into this place, this toxic soil of Empire. I had worried that it was madness, the compulsion to repeat unresolved traumas of the past, rather than the clear voice of intuitive knowing of the rightness of this place. I had doubted what felt like the profound calling to leap, and let ourselves fall onto the garden's soft bosom.

But in the shadows of winter, and the clear light of spring, this garden had showed me to myself. I had found all parts of myself in its corners, connections for so long unimaginable in what had seemed such a hostile land. But when I had space to begin to get to know it, the land itself – the clay, the plants, the very contours of the earth beneath me – had welcomed me. And in the clumsy thrusting of my hands over and over again into its receptive soil, the garden had responded with this orgiastic frenzy of blossom, my creative potency renewed. This garden had opened itself to me, had seemed to relish my care and bloom in the face of my love. And, by blessing me with this lushly abundant beauty, had seemed to love me right back.

It seems clear now. This is the place where we shall grow old and learn to find contentment. I let joy spill up and overwhelm me, and as I sink into belief that we have found our proper home in this world, tears wash my face and water the plants at my feet. I kneel to touch them, a prayer of gratitude to this place that has so tenderly caught our fall when we leapt in faith. Together, we shall thrive.

My phone pings. *Have you seen the news? Are you OK?* Ping, and ping, and ping again. I read the messages, open a news app, and then social media sites, one after the other, uncomprehending of what I am seeing. A Black man lies on the ground, a White police officer kneeling on his neck. He says he can't breathe, again and again. He calls out for his mother, a mother. He takes his last breath and my screen goes black.

I stare unseeingly at the garden around me. I have been punched in the chest. The atmosphere around me has disappeared, sucked into a black void from which I cannot escape. No matter how hard I try, I find myself in the same position again and again. This garden is no different, its black skin cannot protect mine. Everything has changed, but still I am trapped in the same old cycles of pain. I gasp, and gasp again, as bile rises. It feels as though I cannot breathe.

Blossoming

Rose

MY GARDEN HEAVES and groans with flowers. Stems bow beneath their weight, pollen dusts the ground, bees dip and swerve drunk on nectar. Every colour is here, nothing is curated. It is a fusion clash of brights and pastels, fuchsia penstemon next to gentian veronica near to pastel yellow phlomis close to pink mallow tree. The loudly competing flowers launch a mass offence on aesthetics. It is a carnival of plant life on these English terraces. The plants dance and jostle down the garden in bacchanalian display.

The experience is no less varied with eyes closed. Open the front door to meet lavender buds just outside, stretching over to brush against legs as you walk past, an invitation to crush them between your fingers and carry their scent with you. Chamomile lawn spills out onto the main garden path, and the trodden scent lingers to meet the rosemary. Herbs line the sleeper steps to nowhere: thyme, chives, salad burnet, curry plant, mace and mint rise to meet you as you walk past. A topiary bay at the top to rub the leathery leaves of, then the turn along the perimeter path, lined by dense clumps of lemon balm. Follow the path down to where the wild roses and honey-suckle are in bloom in the native hedge, their scent rolling heavy over all the rest in the warm, long evenings. A wander through the garden is an assault on delicate senses.

Not that I can smell it. It is peak hay fever season, and with every venture into the garden the grass flowers from the fields beyond assault me more strongly. I retreat inside, nose and eyes streaming. I sneeze, and cry, and cough, and wheeze, and cry, and cry again. I am allergic to the best of the English summer.

I can hardly tell the cause of my tears. The news leaves my soul as ragged raw as arms flayed by the briar rose in an unguarded brush against the native hedge. A planting meant to guard the boundary, to protect, only cuts me open, my blood red on its thorns. The native slices through my foreignness and reminds me that I do not belong. I have lost my boundary.

In this disorienting time of being in a pandemic, attacked by invisible molecules on the very air I breathe, which cause my body to attack itself in turn; attacked by the violence of an online mass resurfacing of trauma and death and oppressive pain that then summons old memories to rise within me in unholy call-and-response, I am one suppurating sore of pain, the source of which I can no longer locate. My eyes stream angrily, and I do not know if it is in response to the grasses, or the memories that keep intruding, or in fierce mourning for all the flayed and drowned and lynched and shot and strangled and dead, so visibly and invisibly. Black bodies. In the loss of my boundary, all those bodies are also mine.

I hide my body indoors. Lockdown is gradually being lifted, we can go out and do more, meet in gardens, mingle in small groups outdoors. But every outing feels like an exposure, my Blackness seeming to mark me against this politely silent green and White backdrop. The quiet of the countryside, eerily enhanced by the strange stillness of lockdown, now containing within its emptiness a menacing sense of threat. The simple walk to the village shop for milk is torture, senses on high alert. The shop carries a handful of newspapers which have been ordered for collection. The headlines of most of the papers on display blare loudly against the 'woke'; there seems no counterpoint. The wording angers me; I wish I did not have to be awake to this pain, could sleep-walk in ignorant bliss though this life, but I am allowed no anaesthetic.

Rage swirls and swoops within me. I am desperate for the garden more than ever. As I see others doing through the small squares on the screen of my phone, I long to immerse myself in its uncomplicated sanctuary, to escape to its pure haven. To ground myself in its cool soil, for the damp clay to draw down some of this dry fire that rages over my itchy skin in waves of hives. I long for the clarity of mind that comes with the unthinkingly repetitive tasks of tending it, to bring relief from the foggy thinking brought on by ingesting too many antihistamines and too much of others' opinions and states of mind online. I venture out when I can: when the pollen levels seem lower in the middle of the day, or when whatever combination of pills and sprays that I have desperately taken have had partial effect. I look at the ongoing clear blue skies, longing for rain. Anything to wash the air clean of this invisible toxin that chokes me.

One day I am sitting just inside the closed French doors,

looking out at the vibrant green of the field beyond. I long to go out and walk through it, climb through the woods to the top of the hill above our house and take in this place to which I am trying to belong on this beautiful summer's day. But I know that I cannot – when I tried to go out a few minutes before, I was overwhelmed with uncontrollable fits of sneezing and coughing, and became so wheezy that I had to use my inhaler. So I have sealed the doors and windows of the house, and myself within it, cut off from the beauty outside. I look at the rest of English life happening through a pane of glass, at some remove, as ever.

***

The viral footage of the death of George Floyd has fallen like a spark on the tinder of collectively heightened emotions. A wildfire of reaction rages across the globe. People burst out of the homes they have been locked into for weeks and months, and take to the streets. Stories burst forth, the ones that have been repressed, suppressed, ignored and shouted down again and again. The pressure cookers of micro- and macro-aggressions, all the actions of intolerance and degradation and contempt that bubble and simmer and fuel the systems of racial injustice in which we are trapped boil over. It is like the ragged veil that has sat across much of the collective psyche has been torn, ripped apart by the explosion of feeling.

I am raw and sore and torn wide open, forced to confront all the injury borne by my soft, Black skin that I have shrugged off, ignored or suppressed to get through the days and years of my life in this place. I am enraged by well-meaning White people whose opinions I read online through the tearful blur of feelings that constantly threaten to overwhelm me. Their initial reactions seem to be a feeble wash of a guilt-ridden deploring of how awful things are in the US buffered by a defensively delusional and mistaken gratitude that we don't have that sort of thing here in the UK – 'at least Black people don't get *killed* by police here in the UK' – where we are all supposedly so much more tolerant and accepting. I seethe with unbridled irritation at the nearest White person in my vicinity: my husband.

One morning I stumble round the garden almost blind with rage, the assault of pollen more endurable than the denial in the online headlines I am fleeing. I desperately need soothing, and crashing aimlessly round the paths I soon find it in a surprise opium poppy. It is in what used to be a bed next to the driveway, which was thoroughly trashed by the heavy machinery used in the heating system installation. So far this year the space has held a few residual, hardy shrubs, large and close enough to the house to have avoided much damage, and a few weeds appearing among the churned, compacted mud.

And now, in June, a poppy. The newly opened white flower is larger than my hand, each paper-thin petal bearing a large black splodge at its base, with a couple of other hairy, alien pods of unopened buds beneath. *Papaver orientale 'Royal Wedding'*. It is in the shade of the hedge, in deeply compacted mud, a spot that seems entirely unlikely for what I think of as a sun- and free-draining-soil-loving plant. Yet here it is, looking perfectly happy, blooming amidst destruction.

Its resilience makes me want to cry. I long for the capacity to stand tall, unapologetically myself among the ruins of the facade of polite acceptance and tolerance in which I have made my life. How apt to find an opium poppy at this moment. I want to daub my senses and my mind in its soporific juices, and forget, and forget, and forget.

I remember.

The jeers and the jibes of fellow students at university. The open hostility of patients on the wards. College tutors and supervisors who declined to protect me. Followed around shops, slurs hurled from passing vans.

All hurtful, but the personal attacks were more easily borne. The real harm came in the form of the institutional: racial profiling in airports that always left me feeling in danger of detention or deportation, banks that would not take my money on my first arrival, leaving me financially vulnerable, and a body that the medical establishment would not show care.

Age fourteen, with a stomach flu, and taken by my parents to an emergency clinic when I became so dehydrated I could not stand, I was examined unchaperoned by a male doctor. My trusting parents took on good faith his request to speak to me alone, but I was horrified when he proceeded to examine me genitally, and press me for details of a sexual history that I simply did not have. It was many years later that I realised that I had not consented, could not consent, being a child, and my parents had never been asked for theirs. I was yet another victim of the rampant sexualisation and adultification of Black female bodies. In my shame, my body swallowed the violation of its border.

When I eventually had my sexual awakening, I sought the pill. The doctor started me on an out-of-favour, high-dose hormonal prescription. Not knowing any better, I took it religiously. My young body began to be ravaged by severe bouts of thrush. Over-the-counter remedies having failed to control the burning spread, I went to the doctor a few times in increasing agony, to be dismissed every time, until I went with seeping patches that had by then spread down my inner thighs. The doctor tutted that they had never seen a case so bad, and muttered about susceptibility to pre-diabetes in 'people like you', and then, without testing my blood sugars to verify his presumptions, merely sent me away terrified about my potential future on insulin, with the unhelpful and insulting advice to try natural yogurt. (Eaten? Applied? I was too upset to ask him to clarify.) It was only when I went to a student sexual health clinic to renew my pills, and the horrified nurse joined the dots between my suffering and my prescription, that there was an end to my ordeal. I wept at being taken seriously.

More than a decade later, when I was carrying my as-yet-unknown son, in the obstetric clinic at the hospital, to which I had been referred as a standard measure because of the common fibroids that had been found in the course of investigations for infertility, I took myself and my body seriously. My body was my child's home. Surprised to be abruptly told that I would not be 'allowed' to be under midwife care, and alarmed by the High Risk sticker slapped on my notes while my attempts to voice my queries were dismissed, I bristled at the denial of informed care, and of consent. Armed this time with the confidence of my medical training, I researched the risks of my particular condition. The risk of any danger was slight, and all to me, with no increased risk to this longed-for baby I was carrying. And compared to the risks of putting my Black body under heavy medical management in this dismissive obstetric setting, it was one I was willing to take. The second MBRRACE-UK report, which showed in black and white that a Black Caribbean woman like me had a starkly increased risk of death while in the care of maternity services, had not yet been published, though the evidence was being compiled at the time. But I had experienced enough not to trust my body to care to which I had not consented.

I was right not to trust. Back in the hospital eighteen months later, being induced for my daughter, I was offered no analgesia. In a breath between the slamming back-to-back contractions that the drip stimulated, I had the realisation that I had not been offered so much as a paracetamol, despite being so loud in my labour that I eventually heard a consultant chide the midwife to 'put this woman out of her misery'. I did not have the capacity to ask for painkillers while caught in the vice of an all-encompassing pain more intense than any before; I gritted my teeth and eventually held my daughter in my arms carried solely on the wild belief that my body could endure it because I had survived labour once before. But once at home, with enough distance from the moment to think, I wondered why my midwife and doctors had not considered offering me analgesia, and if the unconscious assumption of my Black body's reduced sensitivity to pain had obscured their ability to see my distress.

I was one of the lucky ones. I had made it through puberty, through navigating the complications of sexual relationships, through childbirth twice without lasting physical harm. And yet, I had been scarred. The price of survival was high; so much had been gouged out of my petal-soft skin.

I had endured the harm as a patient, but attack came on my medical career too. Way into the processes of applying for both permanent residency and specialty training as a psychiatrist, the Home Office suddenly changed the visa rights and requirements for foreign doctors in the UK. It had been deemed that there were too many of us, and we were stealing jobs that were the rightful property of those who were UK born and bred. Never mind that these unemployed UK doctors did not exist, and were as yet nothing more than a plan to expand medical school places at UK universities. Nevertheless, the pathway I had been following, folders of documents carefully collated and photographed in triplicate, ten-page forms filled out and submitted accompanied by sums that totalled thousands of pounds in processing fees, was scrapped with immediate effect.

My career plans hung in the balance as the hospital HR administrators told us with stricken faces that they could not process our applications under the new rules. Their systems were in chaos from the disruption, as foreign-born doctors occupied a significant proportion of jobs in the then neglected and unpopular Cinderella specialty that was psychiatry. The medical union, the British Medical Association, took the Home Office ruling to court and won a stay of execution – we could temporarily apply for jobs while trying to find other routes to legitimately remain – but the impact of the xenophobic policy hit the specialty of psychiatry hard. It would continue to suffer from a recruitment crisis, and swathes of unfilled medical posts, for at least another decade.

I realised, through the shock and disbelief of British colleagues at what was being done to friends they worked alongside every day, that the Britain that I saw as an immigrant was a face often hidden from the country's native subjects. When I acquired my British passport and swore my fealty to a photograph of the Queen in an absurd ceremony, the face to which I muttered the vow to be a loyal subject was one that looked kindly benign. I began to grasp that, in the creation of its Empire, Britain was able to remain Great on its home island, noble and morally upstanding, the admirable nation who was first to pursue the abolition of slavery despite the great cost to its own interests. In its great expansion, those subjects at home had been allowed to forget how they were at the mercy of its subjugation. That pure, illusory face that they came to know was maintained because so much that was vicious and brutal and vile was projected out into the colonies, onto darker, less worthy bodies. The face that looked down upon us out there was one twisted into a cruel sneer of contempt.

I had just seen those cruel sneers on the footage of crowds marching through London, raising Nazi salutes and chanting racist slogans in angry reaction to the outpouring of rage and grief around the world evoked by the viral footage of the death of George Floyd at the end of the previous month. Something was shifting in the collective psyche as it was changing in my own. The projections that had upheld the fantasy illusions of great British values were being forcibly reincorporated, and the psychic pain of all that had been split off being returned was intolerable. Unbearable rage courses through us all. And while hot tears flood my cheeks, my memory overflows with all the times I had been told to go back where I have come from, and in my mind's eye a body bursts into flames, disintegrates into ashes which float on the wind to be deposited across the seas.

***

I am curled in my favourite nook in the sitting room, cross-legged on the floor, back against the foot-thick stone walls of this oldest part of the house. I am hidden behind the curtains, looking out through the closed French doors onto the trees. Huddled behind the invisible impenetrable layer that always lies between my self and this land, I am not sure what or who I am hiding from, besides the painful feelings that I cannot escape. My mouth is dry. I realise that I am thirsty, and my vision focuses on the small clouds of white flowers floating at the edge of the woods opposite. The elder is in flower, and it is having a good year. I remember that elder is one of the herbal remedies recommended for hay fever. Not being in the garden much recently, I have fallen out of the habit of making my garden teas, but I get up and fetch the teapot, resolute to brave the assault of the garden to harvest some of its medicine which I so badly need.

I walk round the garden feeling my way through the plants and putting into my teapot whatever seems to speak to me. I gather a pink elderflower from the small ornamental plant with dark, deeply feathered leaves near the veg beds, for its supposed benefits in calming the immune system and alleviating hay fever. For the taste, and mental clarity, I pick some mint which has escaped its square and is making a bid for freedom down the herb steps. I identify a neat plant with pretty purple flowers and a delicious smell growing through gaps in the pavers in front of the dilapidated Wendy house as selfheal, and add some to my pot for its self-explanatory healing. I add some lemon balm, because there is a lush patch of it in a shady area tucked away beneath a small tree we have not managed to identify at the back of the garden, and it is supposedly calming, and every bit of my nervous and immune systems need calming.

On my way back to the kitchen through the conservatory, I stop to admire the hot pink penstemon at the back of the bed just outside the conservatory doors. Two months ago this was a mess of tangled brown sticks whose identity I did not know. I cleared and tended the space, and now a mass of brilliant pink flowers dance over their narrow leaves. They blaze forth, covered in the humming of bees. To my mind, in this moment, it feels like a beautiful expression of the garden's rage.

I turn around to go inside, glancing towards the beech hedge, and am surprised by the sight of a single, large, glorious white flower among the folded green leaves. It is a rose, at the point of fully opening, petals soft and tender, vulnerable centre of the lushly double flower fully exposed. Such a contrast to the vicious thorns I can see from here, lined up on the bowed stem to protect the plant from hungry predators. This beloved plant of England, herb best known for its use in aiding love.

On impulse I go back out, climb up onto the narrow bed next to the road and tug at a few of the outer petals. They come away easily in my hand, and then the rest of the flower collapses onto the ground below, covered in a scruffy mix of dry moss, beech leaf-litter and the thinly distributed, small white stars of sweet woodruff. I gather up the rest of the fallen petals and put them into my pot, taking it as a sign of the medicine that my sore heart needs. I notice a couple of other buds, but the plant looks sickly. Before going back indoors to make my tea, I get a bucket of mulch and place it around the base of the rose. It is in a most unsuitable place, probably planted before the hedge, to flower against the black railings buried beneath the foliage. It is the only rose I have yet noticed growing in the garden. It reminds me of my grandmother and the roses she grew in her own garden, and I feel a small glow of pleasure. I make a mental note to move it to a happier spot in the autumn, but, for now, I hope that my offering will nourish it in turn.

The tea is delicious, and soothing. A clear, pale green balm to the raw red of my suffering. I hold the cup in front of my face and let the steam work on my inflamed sinuses. As my breathing calms, so do my thoughts. The herbs and flowers have a gentle bitterness which is cooling in the heat of the day. I breathe into the paradox of a cooling mug of hot tea, and remember my grandmother's frequent injunction again, the sharp taste of the drinks she made for me. 'You need cooling.' I smile at her memory, wonder how she would shake her head and laugh to see me embracing the tart flavours I once hated. By the end of the small pot I feel restored. A sense of settled calm that has evaded me for days. A temporary interlude before the events of the world intrude, and once more rage floats to the surface like the single petal that rose to the top of the pot as the water from the kettle poured in, floating translucent on the hot, dark sea.

***

The ferocity of growth in the garden startles me. Having mostly grown plants in pots in this country, as I gradually begin to spend time outside again – brief forays seeking the contents for my soothing pots of tea – I am astounded to see how many times bigger the little plantlings that I put in the ground a month or two ago are becoming. As are the weeds. The mulching has worked, better even than I had hoped. The wasteland is no more, and the fecundity of the previously dormant soil seed bank both delights and slightly alarms me.

The speed of growth of the emerging weeds threatens to overtake the speed at which I could ever hope to keep them in check, and I realise that I must focus my efforts on one thing if I am to have any success. Most of the plants appearing are annuals like herb robert, pretty things which are easily removed when they intrude on other plants, and which provide vital soil cover in still-too-bare areas. I mostly leave these alone and decide to stay with the bindweed, crawling through the undergrowth of the beech hedge on brief assassination missions every couple of days to remove the new shoots that have inevitably appeared. Sheets of it have already covered a neglected patch of land that sits above our garden, and out of fear that it might similarly take over in our space, and suppress all that tries to grow with it, I do my best to pluck it out by the root.

My mind wanders as I work. I think about all that lives within us as a society, the pernicious, strong weeds of racism, of injustice, that snake and twist and tighten around our loving hearts. As I follow the bindweed underground, I think about how unhelpful it is to suddenly tug at these white roots spreading beneath the soil; the connection snaps, leaving this destructive element that we are trying so hard to excavate only buried more deeply. These weeds strangling our hearts feel the same, frustratingly buried by denial and dissociation at any attempt to forcefully pull them to light. But, like the bindweed, they soon grow again and begin to show themselves. I wonder if we must all be like gardeners, and grasp hold, gently but firmly, kneeling on the soil of our lives to persistently extract the noxious growth without damaging all the good that already lives there.

One afternoon, while having a cup of garden tea and feeling comforted by the work I have been able to get out and do in the garden that day, I am still musing about the bindweed and the things we need to uproot from ourselves when a flurry of activity on my social media feed catches my attention as I browse. Someone has written to the editor of a popular gardening magazine to ask what it will be doing to feature more diversity in the representation of gardening in its publication, and to make people from all walks of life feel welcome. The response shared online stings; they would very much like to appeal to a more diverse audience, but fear that perhaps they are not the 'type of people who garden'.

A hot, white rage rises within me. At first it is mute, incoherent, but as I gaze at the plants and flowers around me, they ground my formless incandescence and almost from the soil itself the words rise cold and clear. I type a post; for the handful of followers I have online, it goes relatively viral, shared dozens of times by prominent gardening figures with much larger followings, liked by hundreds of people. I am enraged because my type of people had their land invaded and stolen because it was rich and fertile land for horticulture. My type of people had their families and languages and loves and lives stripped away, their bodies stolen and shipped across the seas to forcibly farm that land because they were expert gardeners. And after hundreds of years of brutality, my type of people were left, having embraced the colonial horticultural systems – growing neat domestic front yards, tending Royal Botanic Gardens and having membership in local horticultural societies – because that was all that remained. The audacity of an institution which holds itself as a leading light in British gardening to be so wildly ignorant of the past and present on which its existence is founded infuriates me.

I take my rage back into the garden. It can never be a place of uncomplicated sanctuary for me, never be a haven of escape from the pain of the world around me because it carries the history of all the pain that my ancestors went through in relation to the land. And it certainly cannot bring me unadulterated peace while in the present moment key leaders in the gardening community tell themselves Anansi stories to uphold their beliefs about who does and does not belong in the picturesque English garden. This is my picturesque English garden, and I want to attack it in displaced rage.

Walking around looking for a satisfying task to channel the anger I feel roiling within me, I find myself at the back of the razored patch of bamboo. When it was first cut down, there seemed to be only a desert of bamboo sticks left in its place. Some of the bamboo has come back to life, with my husband occasionally tackling a piece that begins shooting for the sky, painstakingly digging them out bit by bit. I do not have the sheer strength to dig out one of the stalks that intermittently wave over this space, growing visibly taller every day, but I am interested in the plants that have sprung back to life in the light made available by the cut bamboo. I squat and look at a line of ferns, yellow loosestrife and poppies, and wonder how long they were buried there, still surviving despite everything this grove of bamboo did to suppress their growth.

I notice a set of pretty, feathery leaves and recognise them as astilbe. It is one of my favourite plants, with its gorgeous plume of blooms, although this one is not even in bud, and looks sickly. I am not sure how I know, but I sense that something is killing this plant, and it needs to be moved if it is to survive. Increasingly, I am trusting my gardening instincts, believing in my ability to understand what the plants are telling me.

I get my hori knife and start to furiously dig the plant out in order to relocate it. Suddenly I feel fiery, angry stings on my ankles as I am bitten once, twice. I look down to see red ants swarming over my boots, carrying small white eggs from the nest they had made within the astilbe's roots. Attacking me as they felt attacked by my flailing attempt to dig up their nest, as I felt attacked by such ignorant, unthinking words.

I haven't been bitten by ants for decades, but the burn of the formic acid inflames a childhood memory of inadvertently standing on an ant nest. I remember being in my grandmother's garden, peering at the roses and wondering why their flowers were supposed to be so special when they were a lot less amazing than the bougainvillea down the road that stopped me in my tracks every time I saw it. Then I felt the hot stings on my legs and realised I had been standing on one of the many shifting nests that the ants seemed to love making among the roses' roots. I jumped about, shaking my legs and whacking at them to try to get the ants off, but the sharp lash of their stings felt like retribution for disobedient thoughts about the revered flowers of the old massas.

Standing up and stamping the ants off my legs now, I look up and the rose in bloom in the hedge opposite catches my eye. I think of my grandmother tending her roses, believing that to grow Englishness was to grow better, believing that her Indigenous roots were something to stay shamefully hidden, and I have a sudden, vicious urge to rip the plant out by the root.

I drop the ant-ridden astilbe on the ground next to me, knowing there is nothing I can do for the plant until the ants have all vacated its roots. They soon move their nest to a new location. The hot, angry welts burn on my legs for days.

***

Summer solstice arrives. The longest day in a year of days that have felt like the longest I have ever known. After the children are finally in bed, I wander round the garden. It is at its peak of fertility and beauty, which seems to taunt me. I am spent. Exhausted by weeks of allergies, and the flood of rage-induced cortisol and adrenaline, my immune and endocrine systems' overdrive has left me feeling ill. Although whether it is simply that pollen levels are finally abating, perhaps subdued by the rains that have eventually come, or due to all the garden tea I am once again drinking, the worst of my allergic reaction seems to have settled. But my burnt-out weariness seems completely out of tune with the cycle in the garden around me. What had previously been a pile of brown sticks next to a small well carved out of the stream has become a pink and white sparkling mound of *Erigeron karvinskianus*, cascading over to form a soft skirt dressing the stone edge of the stream. Some furry clumps of leaves on the other side of the stream now reach to the sky as towering stands of foxgloves. Uninspiring dead brown patches have transformed into mounds of leaves held up on thin stems, glowing purple flowers held above them. Geraniums, which I always thought of as rather boring, but here looking magnificent, and none of it reflecting the way I feel.

But then my breathing slows and my racing mind settles enough to allow me to start to notice the plants which had flowered in the spring and early summer. The helle-bores and primroses, a bell campion just above the emergence of the stream, which had been covered in a full flush of fat white flowers just a few weeks before, echoed at the time by the white star-like flowers of snow-in-summer, which had bloomed above silvery leaves just outside the conservatory doors. The spikes of purple veronica which reappeared after my hours spent clearing the thicket of ornamental raspberry cane that was over-whelming their bed. Now those plants looked tired and worn out. Yellowing carcasses of faded petals and drying seed heads are held above shabby leaves. These were the plants that mirrored my internal state, and it comforted me to notice them, and realise that not everything was at its best in the fullness of summer.

I had read an article recently which had also comforted me. It was about a phenomenon that seemed the opposite of Seasonal Affective Disorder as we traditionally thought of it, describing a population of people who felt at their worst at summer's peak. It seemed to disproportionately affect those from tropical climes who had emigrated to the Northern Hemisphere, and was proposed to be a response to the changes in extremes of daylight that were not a feature of life at the equator.

In the reading so much made sense to me. Others always assumed that the summer months, with all their heat and light, would be my favourite, and yet it was a time of year that discomfited me. My sleep was troubled by the long days, leaving me tired and worn rather than energised by the light. My favourite times of year, despite the cooler temperatures, were the transitional seasons of autumn and spring. That those seasons contained the equinoxes when equal-length day and night matched the environment in which I was born brought a level of self-understanding which was a relief. I no longer had to wonder why this time of year that I, along with everyone else, assumed I should love was not my favourite. I was not weird, or broken. There was a reason, wired into my brain by the very sun.

The relief is freeing. I remember that I am no longer confined only to the garden; restrictions have lifted and we can go out for more than our once-a-day walk now. My son has even returned, somewhat hesitantly, to school. I kept him home for the first week that the village school reopened, wary of how he would cope with social distancing, and masked and visored teachers, until one of his school friends passed by on a walk, saw us out in the garden and begged him to come back and play. I relented, moved by the little boy's pleas and by the thought that a bit more space might be good for us all. Settling back into the still-surreal rhythms of distanced drop-offs and class bubbles has been going unevenly, and handling the emotional fallout of this has no doubt added to my exhaustion.

I take a walk through the quiet village. I encounter no other people as I make my way up the main street, notice no other creatures. It is just me and flowers, and I feel my guard soften and lower in their presence. I slowly

walk past and admire front gardens, stopping to photograph and smell the roses that spill over fences and tumble
down the front of buildings. Quintessentially English,
universally beautiful. There is a stretch of small cottages
near the middle of the village without any front gardens
as such, but where plants bloom on the pavement wildly
anyway. It is the most colourful stretch of the village's
high street. A mixture of wildflowers and cultivated
escapees from nearby gardens have made themselves at
home in the cracks on the pavement and in the walls, and
they jostle together and flower profusely in this unlikely
space. There are cheerful yellow and white ox-eye daisies
and tall, bright blue chicory, alongside lime green
euphorbia, a climbing rose, buddleia and towering hollyhocks. The latter particularly please me; something about
the shape of the petals and stamens reminds me of hibiscus
flowers. I stop and look at them boldly thriving in the
places they have made for themselves right in the centre
of this English village, and wonder what might be possible
for myself.

A little further along I am stopped in my tracks for
another reason. A fellow villager has painted their gate
in the colours of the rainbow, the symbol that has been
co-opted in the last few months as one of gratitude for
the sacrifices of healthcare workers in the NHS. But now
over the rainbow has appeared a large sign in black and
white, speaking loudly into the surrounding quiet: BLACK
LIVES MATTER.

I am overcome. Isolated in this new place, I had projected
my rage at all the past hurt I had borne onto this stereotypical English village, easily done when the honey stone

stood impassively silent. My sweeping assumptions were wrong. I feel vulnerable, a lifetime's defences shattered.

I turn and walk home, stumbling back past the hibiscus-hollyhocks, unseeing through my tears.

Crocosmia

A FEW CLUSTERS of strappy, accordion-folded leaves have appeared in the garden. They lengthened over the spring months, but now, as the hot July days stretch out, thicker structures appear from within them. At their tips rows of bumps form, looking like teeth. They stretch and grow into zipper rows of buds that colour from their starting pale yellow through orange to finally bright red as at last they reveal themselves fully. Arcs of flowers held up on the end of strap-like stems opening sequentially from tail to tip, a deep blood red. One of the first to open hovers over the steps down to the patio. The red flowers hang suspended over our heads like ruby-throated hummingbirds.

Having that thought makes my throat thicken and eyes begin to sting. I miss hummingbirds. The last time I saw them hovering overhead in numbers as thick as the crocosmia flowers floating over the patio was in Trinidad, and Trinidad has never felt so far away.

After my parents sold the house that I had thought of as our family home, they bought a place in the east of the island. It was in a neighbourhood unfamiliar to me, but on our first visit to their new home, in our search for places to go and things to do with the children that did not require making the long journey up the highway back towards Port-of-Spain, we discovered a magical garden nearby.

Cafe Mariposa is in the Lopinot hills, an area that had been renamed for the French count who built his plantation estate and settled in the area in 1805 while the island was under British rule. For a long while Lopinot was inhabited by the family and enslaved people who made up the count's household, and other French creole migrants who joined them, attracted by the rich cocoa estates the land nurtured. The settlement, centred around the La Reconnaissance estate, remained largely untouched by change until the British government took ownership of the land in the early 1940s, and populated the village with people they displaced from the region of Caura.

Caura had been one of the island's largest and most prosperous villages, despite its remote location in the northeastern hills. It was a community in a fertile valley of the Northern Range where the patois language and culture of the island's mix of Indigenous, Spanish, French and African people and influences thrived away from British oversight. It had a school – rare for such a remote location – and one of the lowest infant mortality rates in the country until the 1940s, when the government under governor Sir Bede Clifford decided to dam the Caura River and build a reservoir over the site of the village to provide water to the entire north of the island. More than one thousand residents were evicted, and their homes dynamited and destroyed. Many of the old Caura villagers were relocated, travelling on foot with their children and what few possessions they could carry on their backs to the

Lopinot valley. However, the project for which all those lives were put through forced upheaval never came to pass. It was rife with corruption and bribery involving the governor and many expatriate British staff brought in to serve as 'advisers' and 'experts'. By the time the next governor, Sir John Shaw, declared the project abandoned a few years later, the previously thriving community of Caura was a gravesite of gaping excavations and expensive machinery rusting among the returning tropical rainforest. It is still considered by some to be the biggest financial and social scandal that Trinidad has ever seen.

My head full of hummingbirds and the stories of the Lopinot and Caura valleys, I wander through our valley to the bottom of our garden. We were finally able to get a contractor back to fill in the trench that has gaped all this time on what used to be the lawn. The ground has been infilled, new drainage pipes laid to help with the flooding, the site roughly levelled. We also got them to dig out a small pond and widen the stream to increase its capacity to better cope with future winter floods. They have finished and are gone now, but the site is a wreck. The carefully placed stones that edged the stream before are torn up and muddy, and half of the gravel of the driveway has been lodged into the soil. It pains me to look at the destruction we wrought on this site that so charmed us when we first arrived. I cannot see how yet, but I hope that in time this landscape will regain some of the beauty that we met.

Lopinot village retains traces of the culture that was brought from Caura with the new residents. It is one of the pockets where patois can still be heard on the island,

and some of the old agricultural and culinary traditions remain. It is said to be haunted by a soucouyant, and certainly there is a melancholy that drips with the mosses and bromeliads from the wide spreading branches of the samaan trees that front the Lopinot Historical Complex. A grief that can still be felt when descendants of the Caura migration, such as the Guerrero sisters who run Cafe Mariposa, speak of how their family came to be there.

In their garden, these sisters have composted grief into beauty. The restaurant serves a daily-changing menu based around the heritage cocoa – the Trinitario bean, a hybrid unique to the island – which is still grown in the valley. The balcony on which we dined was lined with feeders full of the nectar that hummingbirds seek, and the thrum of their wings accompanied the meal, along with the popular harmonies sung by the choir of sisters. The balcony overlooked the garden, which was planted with traditional herbal and medicinal plants, many of them lost to common memory. After eating, we walked through the garden and up the trail into the rainforest on the hills above; a walk that bridges through time, simultaneously in the present Trinidad, and in the Iëre of hundreds of years past. The home of the original people, traces of whom can still be seen in the features of the women who run the place, who had received us like family.

We will not be visiting their beautiful garden in Trinidad any time soon, and the grief of that haunts me. The UK is somewhat unevenly opening up, with a drive for the reopening of shops and restaurants while locally enforced lockdowns remain an uneasy reminder that the pandemic is not over. However, Trinidad and Tobago remains tightly shut against the virus. The country has virtually eliminated Covid and avoided community trans-mission. I read a small article in a specialist journal about how the Trinidad and Tobago government's response has been ranked number one in the world according to WHO criteria. But while New Zealand is lauded in hundreds of newspaper articles, Trinidad's similarly incredible efforts to keep its population safe go darkly unremarked.

I sit on the curved steps that lead down to the mudscape at the bottom of the garden, under a patch of crocosmia arching beside a red-leaved acer that tumbles down beside the steps. From this vantage point of both comfort – the crocosmia hovering at my back – and unease at the muddy destruction we have enacted on the garden, I video-call my mother. She had back surgery at the end of last year, and I am concerned about how her progress will be affected by the ongoing loss of her physio and rehab sessions. I peer at her down my phone screen, trying to assess how she is doing, and fretting that I cannot see her in the flesh. I have never felt so cut off from my family before.

When Oli and I made the decision to settle permanently in the UK, it was with the proviso of regular trips back to Trinidad, for the grounding in my childhood home that I need, annual doses of food, sunshine, and something else that fattens my soul. A combination of being surrounded by black and brown skin of all shades and immersion in the local dialect that nourishes me even more than the curried goat roti with an ice-cold Carib that was always waiting for me on touchdown.

On the descent to Piarco International Airport, as the

outline of the hills of the Northern Range come into view, a landscape to which I am imprinted like a duckling on the first thing it sees, I feel something within me sigh and loosen. By the time I have landed on home ground my body has changed, chin raised, shoulders dropped, hips eased and voice box adjusted. I slip from the more clipped English tones that I have adopted over the years. At first consciously, to counter all the patients who could not or would not understand me, and to protect myself from long discussions about where I was from that too often seemed to end in some tale about a coloured man they once knew or an irrelevant story of a violent crime that had happened on any, it did not matter which, Caribbean island. But with time my speech changed more insidiously, reflecting the way my mind has been shaped as I began to think in British English. Once back on home ground my vowels lengthen and sing, and my thought patterns change in subtle and unsubtle ways, and I become, once again, a fully embodied Trinidadian. Without access to that reinfusion of life, I feel a haunted, pale shell of myself. No longer entirely Trinidadian, now that I have lived here longer than I lived there, but never fully accepted as British, a half-formed half-caste set adrift by the pandemic on this island.

My eyes blur with tears as I catch sight of the brightly tropical *Crocosmia 'Lucifer'* reflected in the image of my mother on my phone. They look heavenly, soaring over the garden while I feel trapped in this hell. The deep sadness of missing my childhood home rises up, through and over me in waves that I have not felt in this strength for years. I am desperate for home. The tropical-seeming flowers on which my eyes are fixed, like an anchor to

keep me stable through this emotional storm, remind me
of the glasshouses full of tropical plants in my university
town that I visited when I felt most lost at sea. They were
beautiful buildings, full of comfortingly familiar plants
uprooted from distant lands to be grown here, in glass
boxes. The glass clouded and streaked with condensation
from the warm, humid air inside. Melancholy dripped
through the spaces with the plants' guttation. And yet
something of the feeling of that familiar warm humidity
eased the bruising on my brown skin from trying to make
a life for myself here, efforts at creating happiness that
seemed to be continually spoiled no matter how hard I
worked, how much I tried.

***

Last year we had returned to my childhood home, not for
the usual annual warming of our bones and recentering
this part of my identity, but with a speculative eye to a
permanent return. We avoided my favourite beaches
cloaked in tropical rainforest, and spent time in cars, on
the roads, navigating our likely daily commutes, surrounded
by concrete. We considered the high cost of living, inflated
by oil reserves. For all our years overseas earning a currency
that had been valued far more than the Trinidadian dollar,
it seemed that the same property bubble that had made
my parents' sale of our old home so financially advanta-
geous ironically meant that we could hardly afford a home
in the valleys that I loved best, their sides carved out ever
further, urban development spreading like the increasing
wildfires over the hills.

A return would be a challenge, but a challenge that was not insurmountable. So when I floated the idea of taking a career break upon relocation, time to replenish my burnt-out resources and help our family to ground into different soil, I was instead shown a job advert. One for a newly created position due to start soon after our proposed arrival. One to reimagine and restructure mental health care on the islands – an epic task. One that would leave no room for secure grounding. Should I apply for this position, as it seemed I would be encouraged and expected to do, I would have to hit the ground running, soar into flight before I had found my feet. My heart fell at the discordance between my imagined return and the one lovingly envisioned for me.

We met potential future colleagues at the hospitals. They were brutally honest about the uncertain welcome we would receive, perceived as British intruders coming to lord it over the locals. The expectations of my success had grown in my absence. On my arrival they would be colossal, and weighty, but I would have no prodigal daughter's embracing return; in the land of my birth I would be seen as a corrupted foreigner. The native turned invasive.

Troubled by the shape that the landscape of home-coming was taking for me, I went to lime with old school friends, and paid close attention to how their children were growing up. Safely locked away in gated communities from the ever-growing levels of violent crime, safely sealed into air-conditioned cars and houses from the ever-intensifying heat. We would be boxed into a life near enough to the major hospitals for us to work, safe enough to protect us from the risks we would attract as unwilling expats. Seeing it clearly was heartbreaking; a realisation that, as I had changed, while I was not taking heed so had life on my beautiful island. It seemed the wounds of colonialism that lay in my relationship with home had only gaped wider over time. If I wanted to give myself up to the soil that had birthed me in order to heal the pain of my own wombline, it felt that I could not do so on my mother island.

Then we visited this house in a village where everyone was known, and the children ran from house to house and played in the river and trees, where my own children could run freely through the woods and fields. It felt more familiar than anything I had seen in the place I still thought of as home. I saw the possibility of all the best parts of my Caribbean childhood in this English garden.

I wanted to give my children what I half-wished, half-remembered. A living landscape that held them, a mother in every neighbourhood home, as well as every tree and curve of the land. Somewhere they were intimately known. I wanted to give them a place where their belonging was complete and unquestioned and mutual. I wanted them to have a home. And like my childhood steeped in memories of the outdoors, I wanted them to have a verdant childhood in a garden.

But that garden would not be tropical. Following a desire to root down among the plants led to a difficult conclusion – at this point in our lives, it seemed that we could raise our children and shape our lives with the most freedom in the UK. After twenty years the long-stretched umbilical cord that had held on in the one-day-maybe dream of a return to tropical home soil was cut at last. I landed in the hollow of this garden a tender, bleeding, infant thing.

***

I am still deeply missing my childhood landscape, wistful and nostalgic for the gardens I grew up in. I remember sitting on the green-painted back steps, a blaze of Antigua heath cascading down beside them. Our cat would hide among the flowers and try to hunt the birds that came to feed on the nectar. Big-eye grieves would sing their warning call, dive-bomb her hiding place until she slunk away. I remember the huge gmelina, long racemes of maroon and yellow cascading over the gate at the side of the driveway. Next to it the datura, the devil's trumpet. Large, creamy-white, bell-shaped flowers with their drowsily heavy scent laying thick over the bottom of the garden every evening. Its touch was poison, a deadly beauty.

Other night-scented beauties gave life. The jasmine, flowers shining like stars, whose leaves were used for tea. And my father's favourite, the lime hedge from which he made lime-bud tea; it flowered sweetly just before the rains. Evenings also brought the bats to the calabash tree. They pollinated the night-blooming flowers, creating the large gourds that would follow as fruit. And at the back of the house a bank of ginger lilies growing as high as the roof, brightly marked spiders living among them. The house on many levels, at ground level on one side, one floor up on the other, a basement level carved into the steeply sloping hillside in which it was set. That home and garden no longer exist, sold and demolished to build a huge new house occupying almost the entire plot. Even if I could fly back to Trinidad again, I can never go back home. My heart keens at the truth of it.

A friend online recommends that I read Jamaica Kincaid's *My Garden (Book):*. I search out a copy online – it is not easy to get hold of, seeming not to be currently in print in the UK. When it arrives I immerse myself in it like she immerses herself in the bath with a seed catalogue. It is a delight to read, and I chuckle along with her observations of being someone from the Caribbean gardening in a temperate place. She was born in Antigua and the book largely deals with her experience of making and tending her garden in Vermont. I read along with deep pleasure, and then one chapter, 'What Joseph Banks Wrought', punches me in the gut.

There on a page she lists out common plants of the Antiguan garden, also common to the gardens of my childhood, and notes that none of them belong to the Caribbean at all. Bougainvillea, plumbago, croton, hibiscus, allamanda, poinsettia, bird of paradise, flamboyant . . . Relentlessly she goes on and on, naming the flowers I have loved and missed, flowers of my grandmother's and mother's gardens. They were all imported under colonialism. She asks who now might know what the island looked like before all that? No one knows, there is no memory of it.

I am devastated. I realise that I do not know my island at all. That all that I am homesick for is a colonial creation; I have been raised to love a lie. The meaning of that love has been annihilated. It was not only the roses of my grandmother's garden that were alien imports; none of her flowers belonged there. Her great-grandmother, the one who was rumoured to be native to the Caribbean, would have viewed the garden I grew up in as foreign territory. I have been rendered even more profoundly homeless in the world, the comfort of the gardens of my childhood proven false.

Blindsided by grief, I go out and walk agitatedly through the woods and up along the path that climbs the hill above the village. All of this landscape is actively managed and maintained, and I realise as I walk that even here, in the English countryside, this landscape that I am gazing on is a colonial creation. Indigenous ancestors of this place would view my garden here, these neatly hedged and divided fields, as foreign territory. It strikes me that the only truly wild places free from human curation and management are the weedy verges, just as it was in the city. This country landscape is no more free from the scars of colonialism than I am.

As I walk I remember my favourite places to visit back home, our family trips to the remote northeast coastline, where fishing was the main trade and the land was never cultivated. I think about trips into the interior of Tobago, guided hikes into the rainforest of the Main Ridge Forest Reserve, the oldest conservation forest in the world. That landscape was conserved after the arrival of the wave of colonisers so set on bending landscapes and people to their will, but those pockets of remote land perhaps offer the last clues to what the island might have once been. I understand why I have loved them the most. I get to the top of the hill and turn back to look at the view over the valley, the village's church spire rising above the trees. To my wetly colonised eyes, it is beautiful.

***

July. I am in the garden with my daughter. It is her birthday, so often the hottest day of the year. Last year we went to the air-conditioned cinema and had a frozen dessert as our cake. This year, for her fourth birthday, a party is not possible, but I have been trying to make the day special for her. We take ice creams with us into the garden and seek shelter. She crawls into the space under the arching crocosmia by the stream, too shaded by the dense leaves for much else to grow beneath them. Longing for her natural ease and comfort in the space, I follow her lead, curl my body and make myself as small as I can, and join her. We dip our feet in the stream to keep cool.

Singing to herself, she is soon up and about again, a joyful energy propelling her. She goes to a clump of Japanese anemone which has just come into bloom on the other side of the path, starts plucking flowers and returning to put them in my afro. Laughing, I put blooms in the loose halo of curls around her face too. Soon we are crowned with blossom. I take photos of us on my phone, giggling, covering each other's faces with kisses as soft as the petals showering from our hair. When I look back at the photos later, I see a beam of sunlight streaming through the hedge behind us. We are surrounded by glowing halos of flowers and light. The garden's blessing upon our heads.

***

I have always raised my head to be seen. It has often felt as though my very presence in the spaces I have occupied in my life in England has been a form of protest and activism, so often have I found myself the only Black person in the room. At university, a naive teenager, I threw myself fully into the experience of the place; if I was to be the only one, I would be shining, at the balls, and the seven-course dinners. The white-gloved serving staff would serve me the fine wines from the college cellars too, and I would enjoy it. At one such black-tie dinner in a candlelit great hall, the conversation around me turned to ancestry and heritage. Warmed and emboldened by a few glasses of wine, I held forth on the stories of my ancestors, what little I then knew of the Portuguese man who had founded the island's first taxi service and married a Chinese immigrant, and the ex-enslaved carpenter who went blind but still worked with his hands, still guided his feet across seven streams to get from one rural village to the next to sell his wares. To my surprise, my tales held fellow diners rapt, engaged with my stories despite themselves. It was a telling reminder that my story should hold space with those of lords and ladies, and distant cousins to the crown.

And then again, at each stage of my training. Living and working outside London, I found myself over and over the only Black person in my training cohort, or the only Black doctor on my team. It was a singularity that I felt, but the significance of which I did not appreciate fully until my time in psychiatry.

The most powerful experience I had of the possible meaning of my particular, Black Caribbean presence to others in this place was during my time in a therapeutic community (TC) on my first psychotherapy placement. Therapeutic communities are group treatment units for patients with complex and long-standing mental health difficulties. Three days a week we all gathered, staff and patients, and, in as non-hierarchical a structure as possible, attempted to live as a functioning community. There were task groups for doing the shopping, for cooking lunch, for cleaning up. We voted new members in, considered and supported one another's crises, and, sometimes, voted those who were not participating in the community in a therapeutic way out. We played games, laughed, and in large and small groups did the often painful and serious business of therapy together. My role was assisting as a group therapy facilitator under the supervision of the more experienced permanent TC staff and my consultants. But as my first direct experience of therapy, and often unexpectedly confronted by the patients' remarkable insights into my own psyche that I had not yet grasped, it indeed felt like I was having therapy alongside them.

And in this unusual setting, where patients were known more fully and as whole people, so was I more fully and wholly seen by my patients. It was a vulnerable sensation, the loss of the medical persona, and a humbling one, becoming instead a person-as-healer, as we all have potential to be to one another every day. But it was also deeply freeing to laugh loudly at a very funny joke, be laughed at in turn for my ineptitude at rounders, speak honestly about the awfulness of Accident and Emergency departments at three in the morning from all sides.

One of the patients who was already a fairly senior

member of the community when I joined was a Black woman. We didn't speak much directly of her ancestry or mine during her remaining months in the group, but I gathered that she was of Caribbean heritage. I did not want to probe unnecessarily, and was still working out my understanding of boundaries and where they felt comfortable for me in a therapeutic sense, but my Trinidadian background was open knowledge to the group. Yet I was stunned to hear her farewell message to me during the elaborate leaving ritual that was part of the ceremony for successful completion of the programme.

As a temporary and inexperienced staff member in the team, I did not expect to feature prominently as an attachment figure, but she spoke movingly of how seeing a Black Caribbean doctor had influenced her. She had survived an extremely challenging childhood, and she spoke of having never been sure what a healthy Caribbean personality looked like. For all her life to this point, she had believed that all the parts of her that did not fit the White English norms were somehow mad, but in seeing me laugh loudly, gesticulate passionately, steups frequently, and exaggerate enthusiastically, she realised that there were aspects of her that were simply normally Caribbean. She regretted that our time together had been relatively short, which meant that she could not learn more about herself through seeing me, but spoke of her immense gratitude that we had this encounter. I went home later that evening and wept at the thought of the little child who, in the absence of anyone to show her herself, had been led to believe that some of the most joyful, exuberant parts of herself were signs of madness.

And so being me, being vibrantly, thrivingly visible and alive in every room that I occupied had been a sort of activism in itself. But it made existence exhausting. Over the years it had become clear that simply existing in these spaces, simply being who I was every day in room after room, as the one Black person there, was not enough. I could not remain mute on things I observed around me; too much could be projected upon me if I was a mere black, silent screen. I could not only be seen, I had to speak.

It bubbled in my veins, passed down through generations in my DNA. I remembered the stories of my paternal great-grandmother, who was born in Dominica but migrated to Trinidad, where she married, had children and campaigned for universal suffrage. Her daughter organised protests against discriminatory employment practices. I had stood in protests, had spoken in our local council chamber against the closure of children's centres with my son strapped to my chest. Now I stood silently in my beautiful garden. What would I do, granddaughter of these grandmothers? How would I make myself heard?

***

The crocosmia are in their fullest glory for weeks, flying above the garden in their unmissable bright red. They look so tropical, and therefore wildly out of place in my imagined vision of an English country garden. But my ideas of what is tropical and what belongs in an English garden have been profoundly upended. I read that crocosmia are fully hardy here, even escaping from gardens to make themselves at home in the landscapes of the

Hebridean islands. I take photos of them in the garden from every angle, tilting my phone camera this way and that, contorting myself into odd shapes around them, trying to capture this sense of them suspended in flight. I post the photos online and look at them on my screens later. Rubies sewn into the garden's green cloth. Blood Pollock-spattered across the landscape.

I am able to spend more and more time outside again, and find myself there frequently, unthinkingly taking my grief into the garden. The busyness of spring tasks seems to have slowed. There is watering to be done, of the house-plants which I place out on the patio for a summer vacation, and the annual veg. For the garden, I take advice that I see online to water the plantings in the veg beds thoroughly but relatively infrequently to encourage the growth of deep roots. Ornamental perennials I have watered-in on planting, dipping the watering cans into the pools formed along the stream's course, and then left them largely to their own devices. The rains have returned, and there seems to be plenty of water held in the clay, and I am hopeful that the mulch, and all the creatures and worms that have followed it, will help make the nutrients trapped in the soil available to the growing plants that need it. I am doing my best to help the earth of this garden become as healthy a soil as possible. In doing so I am trusting that the plants will be given the resilience they need to thrive.

The biggest call I feel from the garden now is to move languorously along the paths and enjoy it. I step outside and wander around the beds, listening for the call to do, to tend, but there is little. The midsummer stillness feels oddly like midwinter, a pause, this time at the top of the expansive inhalation of the year before the swooping exhale into eventual darkness. We are near the peak of growth. The garden is tumbling well out of my control, its lush growth reaching up, out and over, beyond any limits I could have imagined. In places I tie things in to makeshift supports, where they begin to encroach on the path so much that we cannot pass, or collapse onto and inhibit their neighbours. I add 'supports – rusty metal?' to my list of garden desires, although this feels more like a need. The sheer abundance of it overwhelms me. I have never received such a generous gift.

There are so many flowers, but I read that to have even more I must cut them, and delay the plants' going to seed. Something about that line of thinking feels cruel, but I spend an evening with a glass of wine, after the children have gone to bed, deadheading the spent flowers of the purple daisy, whose shining bank of blossom lights up the conservatory. The repeated cuts are cathartic, and I am rewarded later with a fresh flush of bloom. It feels less callous to think of sharing the prolific blossom so generously offered. I pick posy after posy, and while they are beautiful in the house, I especially enjoy leaving them on the doorsteps of others to return their kindnesses. Generosity multiplying many times over.

As I feel the garden grow well beyond anything of my doing, it is disturbing to realise that I have no true control over this space. I might cut it and trim it and prune it to create that illusion, but I begin to feel the deep reality that these plants are beings with a life and a will entirely their own. In the bare earth of spring, I planted two *Thalictrum delavayi* side by side. One of them is towering upwards, covered in buds and soon to flower. The other has disappeared, seemingly dead. I am coming to realise that while I may think of it now as my garden, I am at best a co-creator in this space. My planting is mere sugges- tion; the plants interact with each other and this unique place to grow new possibilities beyond my limited human imagining. I am just one more creature who might grow, maybe even bloom, in this fertile space.

The garden's explosion of growth feels unimaginably bounteous. I feel like I have never seen plants grow so much before, and the beauty of it, so freely offered by all these immigrant beings thriving on this English soil, touches my already tender heart. How did I never know that life could be this effortlessly abundant? I remember the effort of all the hours of patient mulching, but even that had its ample immediate rewards. I am amazed over and over again, and in the vulnerability of my grief I feel the garden's generosity fill my heart. It swells and swells until it bursts the walls that have guarded it for so long, like a seed cracking its case at germination. I bend and touch the ground. The earth is crumbly, friable, where I have tended it. It seems a good place to root.

Verbena

A GROUP OF spindly plants has been rising from the side of the stream all year. Some instinct has told me that they aren't weeds, and I have with some interest watched the thin, square, sturdy stems, with long-spaced pairs of narrow leaves, grow ever taller. Now their tops explode into small afros of vivid purple, the pompoms cheering from the back of the beds. When I scrabble through the plants across the stream to peer more closely, I see that each puff is made of a cluster of tiny, ultraviolet flowers, over which a cloud of pollinators fight.

Verbena bonariensis, leggy beauties hovering over the rest of the August garden. They look sweet towering among the other plants in this garden of mine, but the name leaves a bitter association hovering on the tip of my tongue. This is not the verbena I know from my youth, but a search on the internet tells me that the similarity of names is no coincidence, as the plants are cousins. The one from my childhood, *Stachytarpheta jamaicensis*, or blue vervain/vervine as it was known, is a member of the *Verbenaceae* family, which is native to the Caribbean, and common as a weed. It was also renowned as a herbal remedy for almost every ailment. Just thinking of the plant sends my salivary glands into overdrive as I remember the blue vervain tea my father drank in the hope of helping his hypertension.

I do not know how effective his brews were, and whether they assisted the packets of pills he also needed. He was diagnosed with hypertension in his twenties, and I wonder about how the stress of being a young Black immigrant trying to make his way in New York City affected his lifelong health. I think of how my cortisol has no doubt surged in the last year, and wonder about the state of my own blood pressure after twenty years on English soil. It is no surprise that, no matter where we live, our health outcomes are poorer than those of our White neighbours. Standing here among the plants at the back of the border, moss underfoot by the stream, vivid purple of the verbena at my eyeline, I feel how hot and sore my heart is from the threshing it has taken this summer. And yet it is not over. As the rage and grief of earlier months are metabolised in my body, take their toll on my cells, something as sour and necessary as bile rises within me.

I have been reading the local newspapers reporting on the lead up to Trinidad and Tobago's August 2020 general election, and they leave a rancid taste in my mouth. The country's two major political parties have long been split along racial lines, the People's National Movement attracting largely African-Trinidadian support, and the United National Congress largely Indian-Trinidadian in appeal, representing the two largest ethnic groups in the country. But this year accusations abound of the campaigning being openly racist, with tensions bubbling over between the ethnic groups online. Even when I shut the laptop and step outside, I still see the toxic words burnt onto the dark insides of my eyelids when I close my eyes. There is no reprieve; these divisions have been lashed into us, and reside within. We are still divided and conquered. We loathe ourselves still.

All summer I have been raging at the forces of colonialism that exist out there, grieving for all the harm that has been enacted before. It would be easy to keep fighting against the monsters out in the world, but a discomfort in my veins knows I cannot only look without, I must also look within. It is literal in my case, my creation brought about through the blood of former slavers violently mingling with those who were enslaved. But even more toxic than anything my body might hold are the thoughts and beliefs that infect my mind, that work to uphold the old massas. I know that I unwillingly carry them in me; they were taught to us all as the foundation of this society in which we exist, which we uphold every day. The contagion seems borne on the very air we breathe. As long as we deny and repress examining the infected parts of our minds, the poisoned abscess of Black inferiority we have all internalised will continue to grow. The inflammation builds until it threatens to rip our minds apart.

I am here because of Empire, and while the ley lines that drew me inexorably to this place in England may have been created against some of my ancestors' wills, I cannot deny that, at least in part, I freely chose to remain here. As my eyes drink in the cool purple tonic of the clusters of flowers before me, I feel my mind calm, and still. Reckoning with horrors that have taken root within my self is a distasteful task, but I know there is no hope of clearing them from the world around me if I do not exorcise them from within. I hope the plants will soothe my fevered mind and clear my thinking enough that I may.

***

My grandmother did not tolerate vervine in her garden. It was viewed by many as a weed. She preferred to grow shrub roses on the edge of the grass in her front yard, like the stereotypical English garden of roses against manicured lawn. They sat uncomfortably among the tropical shrubs that also graced her front garden, sickly companions to the thriving crotons, poinsettia, ixora. But her grass was rough and patchy in the dry season, the roses' leaves yellow in the heat. Mounds appeared in the red soil, with dimpled centres, as ants nested among their roots.

My grandmother was one of seven girls. One of her sisters left Trinidad to come to the UK to work; she would have been part of the Windrush generation. My grandmother adored England, and Englishness, but she did not follow them. A devoted Catholic, she was married at eighteen to my grandfather, and by the time she was my age had nine living children, a neat eighteen months to two years between them.

In my grandmother's glass-fronted cabinet, which proudly displayed her best silverware and crystal, polished to a high shine with the Silvo that left my fingers tea-stained every Christmas, and in the nurturing of her beloved roses, some of her Anglophilia was passed on to me. I absorbed it through the pores of my skin along with the lavender water that she would sometimes dab behind my ears as a special treat. As she sat at her dressing table getting ready in the morning, I breathed it in with the rose-scented powder in the round floral box with gilt lettering, lightly applied with a bow-adorned puff to the skin of her neck and shoulders to soak up the tropical heat. Just as I soaked up all the messages around me that this small tropical island was a stagnating sinkhole from which I must strive and climb and endeavour to escape, and that the refined and dignified, anciently civilised country full of the glories of the Empire – Britain – would save me.

***

The internalisation of the island's worthlessness started young. In my childhood in the eighties, talk radio programming seemed to be a constant stream of people loudly arguing about government corruption and ineptitude. I was just born into an era when our first prime minister, Eric Williams, who had achieved independence from the British, was still in power. But after nearly two decades of unchallenged political reign, the shine on our glowing freedom had tarnished. We sought to forge a future for ourselves free of colonialism, when colonial values were all that we remembered. In this newest phase of national development we were as adolescents, clumsy and uncertain in our growth, full of angst and high emotion. And, like teenagers, we were ambivalent about our relationship to our mother nation, too glad to be free of her, but also in some ways longing for her care, as abusive as it had been.

The queues at the United States, Canadian and British embassies seeking visas for emigration were always convoluted and long. Paths to an imagined better life in the developed world where houses were bigger, cars were newer, and the shops were much more abundantly stocked with nicer things were highly sought after. The first independent Trinidadian government had adopted a socialist policy, but also a closed door to imports, in an attempt to encourage and facilitate local manufacturing and production. Foreign-made items were difficult to come by, and the forbidden is always the most desired. Although socialist policy would be seen as being very much on the left of the political spectrum, there was a sense by the growing middle class that things had become rigid, entrenched and corrupted. My father referenced George Orwell's *Animal Farm*, and there was constant debate that while the plantations may have been shut down, the massas still lived among us, in the guise of our own kin. There was a radical political left who dreamed of a better life on the island itself, but what political theory made its way to me through a mix of overheard lively debates and my father's attempts to ensure my political education seemed always to be based on the unspoken foundational assumption that the country was currently, and perhaps inherently, third-rate.

My father worked in government, and sitting in corners of air-conditioned ministry offices doing my homework after school, I would often overhear people waxing lyrical about Singapore. Singapore had gained independence from the British at the same time as Trinidad, and was a comparably sized nation. It was doing well in this marketplace of capitalist nationhood, clean cities full of gleaming skyscrapers and efficient public transportation. I listened to frustrated adults, Black people like I was, trying to make sense of why our island was not like that one. An idea often bandied around about how Singapore had managed to become a 'proper' place, when we had not, was of their inherent superiority as a race. I overheard adults around me imagine Singaporeans as cleverer, harder-working and more compliant than us, who seemed as lazy and undisciplined as our major city.

Our noisy, dirty capital city, a hodge-podge of colonial frontages and rapidly constructed breeze-block buildings of little architectural merit, serviced by an unregulated band of privately owned taxis and small buses known as maxis – a system which completely disintegrated whenever it not-infrequently rained heavily enough to flood the lower-lying streets of the seafront city – seemed a dump in comparison. Street vendors and stray dogs littered the vestigial Victorian parks and squares in which the islanders felt no pride. The place was not ours, it had not been built for us, and so we hated and resented it. We did not remember that it was our ancestors' life-blood which had raised this city out of the swampy marsh it once was, even if it had been shed unwillingly and soaked up by the rich, red-clay soil in the mosquito-laden heat, sprung from a thousand lashes across their backs.

And so I tried to escape it. I was encouraged and, in

acts of great love and sacrifice, given the tools to navigate the supposedly better life that was dreamed of for me. My parents were not wealthy, but found ways to send me to tennis, piano and ballet, to swimming and sailing, to let me read classic works of literature and learn the names of famous artists. They made sure that I had access to the markers of British colonial middle-class living, so that on my arrival at university in England I could try to fit in. I would be a sophisticated and educated young woman, a member of the aspirational middle class to whom 'the world is your oyster': my father's favourite phrase to me.

And in leaving the island with all these resources that had been poured into my upbringing, I was taking part in the 'brain drain', a river of human wealth flooding out to already-better-off nations. A tax on the country of my birth, which I had implicitly believed we were due because of our sheer inadequacy, our worthlessness a drain on the rest of the world's capital, who had to find the money from their own worthy citizens to give us aid.

The love of home was a confusing mix of patriotic pride and celebration of our incredible diversity, with a bitter aftertaste of shame and self-loathing. A loathing which was crudely and sometimes violently projected outwards, so that, for all the talk of our 'melting pot of cultures', fractured fault lines between the races ran through our country like the geological ones deep underground. And every so often the societal landscape would shift, and rumble, and push to the surface a mountain of internecine hatred, just as the Northern Range had emerged from the depths when continents clashed tens of thousands of years ago.

The hierarchy of the races that had been codified by the British to justify slavery – White at the top, Black at the bottom, the nuanced gradations of colourism in between – still enslaved us. We were penned along these manufactured boundaries, and faithful to our old colonial masters, who lived on within us; we upheld them still. These ideas had been beaten, branded, raped, torn, lashed and broken into our bodies over generations. Every depravation imaginable and those the mind shies away from had been enacted on our tender skin, the purple-black bruising still sore in our psyches. Our fractured minds passed on the tortured hierarchies, only slightly distorted. Club nights in my teenage years segregated between 'Golliwogs', 'Coolies' and Whites, no prices on the door for admission, you paid according to the colour of your skin. The price for entry into the supposedly post-colonial world too high. The idea of being post-anything a colonial lie.

***

I am feeling rather delighted with the garden. The planting I have done and spent hours mulching has filled out beautifully, lush green growth covering the bare soil, a delicate lace of flowers thrown over it all. Butterflies flutter and bees hum and damselflies dart everywhere. On my morning wander through the space, by now a daily ritual, I am euphoric to see the ugliness which so distressed me earlier in the year hidden away. And what joy to see the things I raised from seed come into flower! Strawflowers tower above the beds, their spiky blooms like a crown of thorns. The sweet peas are a dream made palpable. With their ethereal colours and intense scent, I vow to always have them in a summer garden. But my favourite are the cosmos, a galaxy of floral stars shining above the beds, their pink and white blossom filling the house as much as the garden. As all these plants that I brought to the space come into vibrant flower, it is clear how much I have changed the garden. My experiments in planting are in full bloom. But, beautiful as it all appears, it seems that not everything will bear fruit.

We sowed too many squash seeds, and in a bid to get all the seedlings in the ground, and to play with the idea of potager gardening, I had planted a squash right in the middle of the main ornamental bed, between the path and stream. The leaves grow like stepping stones between the frothy waves of the rest of the ornamental planting, with attractively curling tendrils reaching out among them. The larger leaves contrast prettily with the other leaf shapes in the bed: a low cloud of delicate erigeron; tall, waving miscanthus; architectural statice; feathery cosmos; all punctuated by the floating heads of button-like scabious. I post a picture online and it attracts much admiration for its beauty, which feeds my ego.

However, there is little substance here. As summer arrived, and all the trees and shrubs came into leaf and exploded into growth, this part of the garden, which seemed in full sun earlier in the spring when I planted out all the seedlings I had been given or nurtured from seed, is cast into significant shade. The other, less photo-genic squash plants, scrambling over the veg beds or across roughly cleared ground in less attractive parts of the garden, are flowering and fruiting enthusiastically. This plant is beautiful but sterile.

One morning, when I am checking on the growth of the squash fruits, wondering how I might protect them from damage by slugs, it occurs to me that my love of the bed which has become my favourite for its beauty is quite superficial. Most of the plants that have made it into that bed are annuals. This was in part deliberate, because seeds were cheap and we had so much time to sow and nurture them. Watching them grow also helped ground me in the early months of lockdown. I sensed, though, that I would need more time to fully understand the nuances of growing on this site, and to help the compacted clay soil recover. These annuals are pioneer plants, breaking the ground for perennials to come. But part of me has forgotten that, and allowed myself to be wholly taken up in their disposable beauty. The beauty itself nourishes me for this season, but it will not be sustained.

I have fallen into the trap of prioritising disposable style over lasting substance. I am trying to grow beauty, but according to whose eyes? I wipe my brow with muddy hands and close my eyes with a sigh, then open them and look at the garden again, questioningly. I know that this image I see is a construction in my brain of the light entering my eyes, but my perception of it – the feelings associated, what seems beautiful and right – that has all been shaped by the things I have been taught, consciously and unconsciously. The gardens I learned to admire in my childhood were colonial constructions. I look at the beautiful bed I have created, and wonder if I am simply re-creating the same thing here.

I am a Black woman, a child of Empire, tending a garden of her own in the English countryside. Sometimes I sit at the top of the garden looking over the terraces below in lush, abundant bloom, and wonder if the peace the garden brings me might be a form of reparations for what my ancestors went through. But I know it is not so simple, and that as long as the residues of colonial structures have made the path to this garden easier for me than it will be for others, nothing is fairly repaid.

For all the difficulties of my existence here in the UK, I know all too well that certain parts of my embodiment have eased it. I have borne the damaging brunt of structural and institutionalised racism, but I have experienced relatively little in the way of personal racist assault. What a thing to be grateful for! But when listening to darker-skinned friends trading stories, I have sat silent. It has felt as if my lighter skin, together with the assumptions of direct mixed-race parentage that it has often carried, have smoothed my way through the White spaces I have been in.

As have the privileges of class. In the early years of my time in the UK, I could often sense British uncertainty about how to place me within its intricate class hierarchy, confused by the combination of my eloquence and accent. But an Oxbridge education and medical titles eventually decided the matter, and over the years I saw myself come to be perceived as solidly middle-class, with all the privileges that carried. It was not incorrect: I had a middle-class upbringing. Despite my father's birth into poverty, both my parents were university-educated, and both had carefully, lovingly, cultivated for me the childhood that they deemed would give me the best chances in life.

Standing in my English country garden, I feel deep gratitude for all their work that got me here. But deeply embedded racism meant that the colour of that hard work would never be good enough. I would always have to work many times harder than others to prove myself, other forms of privilege forever undermined by my race, whatever the colour of my skin. I find myself thinking of my children, and my stomach churns as I think of all the privileges and disadvantages unwillingly branded onto their skin. I wonder if I could ever free them from internalising this burden of self-hatred.

I cannot answer my own question and feel caught in the swirl of trying to make sense of it. I am trying to break down what I know is the false construction of race to create a liberated reality for my children, but sometimes the magnitude of trying to do so while also trying to survive within a racist society makes the world spin before me. To ground myself, I head up to the top of the garden, to the grass under the ash tree. We had let it grow unchecked all summer, reading that tall grass was better for wildlife than mown lawns, but my husband had given it a hay cut about a week earlier. Dutifully following instructions I have read about the proper way to manage a meadow, I take my rake to tidy the heaps of cut grass that had been left to dry in place, and to clear my thoughts with the repetitive action of my body.

As I begin to tackle the biggest heap, the grass shifts of its own accord and something wriggles angrily out from inside the pile. A sleek, golden-brown body, same colour as my summer-warmed skin, slithers out from under the mown grass to lie just at my feet. It is a slow worm, sheltering in this pile of grass which has grown warm in this open site just beyond the canopy of the ash. I had seen one with the children for the first time just a few days before. We had almost stepped on it as it lay in the middle of the path home from school that cut through the fields and wood, avoiding the road. The children had nudged it, but it had not moved. I had picked it up to find it cold, its reptilian metabolism rendered inert. I warmed it a little between my hands, and it seemed to nuzzle into the crook of my arm, where it stayed for most of the rest of the walk home. Crossing the field to our garden, it dropped from my arm and slipped away into the grass.

I am thrilled to see one here, in my own garden. Its smooth body glistens iridescent in the late summer sunlight. After lying at my feet a moment, it disappears back under the pile of grass clippings with what seems to me like an incensed wriggle.

That's me told. I put down my rake and stand staring at the rough grass around me. The slow worm reminds me that I am not the only inhabitant of this space, nor do I want to be. And again makes me question my percep-tions – whose perceived wisdom dictates that I should rake up this grass? Why do I think the piles of clippings look messy? These heaps of warm cuttings have made a home for the slow worm, and its beauty far outweighs anything I could achieve with my rake. This is my garden; it is not a wilderness, it is a space reliant on my shaping it, filled with plants that have co-evolved to thrive with human input. But it has become so clear to me as the months pass that I am reliant on it too. Our thriving will be mutual. Finding my way to learning how we can both bloom without feeling forever trapped in colonial systems of horticulture that cause both our bodies – the garden and mine – harm will be a complicated journey. There is much for me to forget, and even more to remember.

***

All the work that we have put into the veg patch is finally beginning to show results. The lockdown seed-sowing with the children, anxiously tending infant plants, wondering about hardening-off and last frosts, planting out those small plants that survived our amateur ministrations, and the lugging of watering cans back and forth from the stream to keep them watered through the dry spring, is paying off. Some of our crops have not made it this far. The slugs got the lettuce and pak choi before we could, and there are so many slugs. But when I look at the garden overall, their impact seems minimal, so I resign myself to some losses and leave them to their silvered trails.

The children and I climb the steep beds with a trug and much glee as we pull the very first harvests from the ground and pluck them from stems. The soil seems better than when we put our seedlings in the ground a few months ago. There are earthworms now, and everything feels lighter and more crumbly beneath our searching hands; it looks more brown and less of that sickly grey. The carrots do not seem to be doing well in the clay soil, but the beans are thriving, as are the beetroot. It is too early to expect squash, or potatoes, but the potato plants are lush, and the squash plants in the veg beds are in rampant growth, with flowers and the small, swollen nubs of early fruit at their base. The tomato plants are in pots and look indifferent to their growing situation, but there are flowers. A self-seeded borage has appeared in a gap in the shale that covers the steps between the veg beds, and its blue blooms are beautiful.

I take a picture of the pleasing array once we have put it all in the trug: broad beans, some aphid-covered and others part-slugged, but enough to make a meal for us all; huge perpetual spinach, and some brightly coloured rainbow chard leaves; a bunch of radishes that the children have already started to nibble; a couple of onions; a single beetroot. We will not be ready to give up our weekly trips to the farmers' market any time soon, and my respect for the incredible effort it must take to get a significant yield of crops to market has increased a thousandfold from our experience of growing this tiny harvest. But it is our harvest nevertheless, our time and effort and attention grew these offerings, and now they will nourish us in turn.

The children cannot wait to get the veg to the kitchen. They open some broad beans and gently stroke at the soft pillowy insides. We find a caterpillar curled up in one, and they laugh and wish that they could snug up inside a broad bean pod too. They ask me to slice open the small beetroot, and marvel at the perfect concentric rings inside. The purple juice stains my fingers.

Handling the veg like this, warm from the earth, melts something inside me that has been coldly set against the process of growing our own food all this time. I have gone through the motions, urged on by the children's interest, by the empty beds which needed filling, by the horrible waves of anxiety early in lockdown. I was drawn in against my will by the magic of germination of the seeds, finding myself checking up on the trays that littered much of the conservatory nearly hourly, entirely despite myself. Seeing the small harvest we have eked from all those hours of work, I feel even more clearly now how puny our actions were in the immensity of what we were facing. And yet feeling able to sustain ourselves in some small way feels such a necessary act, the pebble with which David felled Goliath.

Our small harvest has not toppled the pandemic. But I feel a new warmth opening up in me to the act of growing my own food. I am still puzzling over my resistance to this, something so ordinary that humans everywhere have done it for millennia, and yet also so vital, not quite able to understand why hunching over the veg beds, weeding, thinning, watering, has tended to make me so out of sorts. The work on the veg has made me squirm internally, brought up uncomfortable feelings, completely at odds to the way I feel when I am tending the garden purely for the delight of flowers. In a quiet moment, I sit on the edge of the beds and let the idea dance loosely around my mind while watching the children playing. As I idly run the garden soil through my fingers, something clicks. I am astounded that I have not made the connection before, and yet there it is. It is as if when my body bends over the veg beds, takes the shapes and goes through the motions of growing crops to eat, the bodily memories of all the pain my displaced and enslaved ancestors endured on the land surfaces. That old trauma lay unresolved, buried somewhere in my DNA, passed on along my lineage. Suddenly I feel it and remember it.

Sat on the steps between rows of potatoes and beetroots, teepees of runner beans racing to the sky above me, I bow my head to the earth and cry. All spring my tears have soaked into this earth. With the plants, they have called forth this growing yield. The garden has composted my grief into bounty.

***

I wake. It is a beautiful late August morning. I take a swig of coffee en route to the garden. It is the weekend and its sacred joy is playing with flowers. No breakfast yet; the garden nourishes me. Wandering through the freshly wet paths, still in slippers and dressing gown, my senses are replete. Like I used to do with my houseplants when we had no garden, I walk around and study the plants, wondering from which I might take a stem or two to bless my house today. I fill a couple of tiny bottles with small posies. Highly scented sweet peas and honeysuckle, the last green froth of lady's mantle, purple stems of clary sage. Vivid beauty clears my eyes, crushed stems stain my fingers, their scent lifting my spirits as bird calls and the song of the stream soothe my ears. My heart beats its own song in response, one of peace and ease and gratitude. This is a meditation, a prayer.

I rest the posies on the patio table and walk up the stone steps, past the little purple puffs of verbena floating above the beds, drawn to the last of the mallow tree's vivid flowers. It has bloomed a pink cloud over the garden for weeks. And every time I walk past it reminds me so much of the hibiscus in the garden in my parents' house back home that my chest hurts. I stand looking at the mallow tree, underplanted with ferns, and the broad leaves of winter heliotrope, then notice an echo of the verbena's purple in the large clump of what I had taken to be a type of grass next to it. I had come up against this plant when going through the beech hedge chasing bindweed, and been sent away with razored arms from the encounter, each fine but tough blade with a sharp, serrated edge. This plant did not want to be bothered, so I left it alone, there at the back of the border, seemingly thriving but begging to be ignored.

Today, though, a circular cluster of its stems gleam red. I ease my way through the plants to have a closer look. There, in the middle of the clump, surrounded by the bright red spray of leaves, lies a glowing purple heart.

I gaze in awe. I have never seen such a thing, and certainly not in an English garden. With the bright pink mallow above, and the big, lush leaves of winter heliotrope and the ferns surrounding it, I can almost imagine myself in the forests of my childhood. I take a photo, then look online to see if the internet can help me identify it. There it is: *Fascicularia bicolour*, a hardy bromeliad. I know bromeliads; they covered the branches of samaan trees around the Queen's Park Savannah, and colonised the canopy of the forests at home. But here was one – a huge one, glowing red and purple like an alien – in my English country garden.

I feel a rush of warmth in my chest. Every time I despair

at my ability to remain here, to make a home for myself in this country into which my ability to put down roots seems so fraught, the garden seems to show me something different. Again and again it feels as though this soil was made for me, in all the planned and accidental coming together of planting that reminds me of home, that feels so oddly familiar in this strange space. Like all the planned and accidental coming together of all the various people who made me, whose forced and fated migrations all around the world, along the trails laid by the upheavals and shifts of empire, set the path that inevitably led me here. The garden calls me home.

***

It is a moment of great excitement. My husband's parents offer to have the children for a few hours – the first child-care we have had all year. The summer lull in viral infection numbers has meant Oli's highly vulnerable parents feel safe enough to brave contact with us. My father-in-law's chemotherapy has been progressing after an initial delay, and we have all been painfully aware of his immunosuppressed status. The irony of our moving nearer to be able to spend more time with family, only to be promptly and profoundly divided from them, has been poignantly felt.

But that is all being briefly put aside. Everyone is extremely excited, but perhaps none more so than Oli and me. We repeat daily how lucky we are to have moved when we did, for me to have timed my career break so precisely right. We see in our children how relatively well we have managed to bear the last few months. But even with all our privilege and good fortune, the months of anxiety combined with unrelieved childcare in isolation of communal support has taken a huge toll. It takes a village, and we have all been cut off from one.

After delivering the sanitised children to my in-laws' garden, we sit in the car and look at each other. It feels suddenly awkward, this intimacy that we have not had for so many months. I am at a loss as to what we should do, but Oli smiles and starts the car. He has a plan.

I sit back and let my gaze drift over the late summer countryside. Eventually, he turns off a roundabout and down a narrow lane. It is a single track with high hedges on both sides, and I nervously hope that we don't meet another car along the way. But soon enough we are making a sharp turn into a rough car park, and there is the sign above the gates: Special Plants.

He has brought me to a garden and plant nursery not far from our home. I remember the roundabout we turned off at, and how before the pandemic we had driven past on the way to my in-laws' house and peered at the intriguing sign tacked up on one side of it: Special Plants, with an arrow. What special plants? What made them special? Without telling me, he had looked the place up and here we were to find out.

We pay the entry fee and go into the garden. I read a small blurb about the nursery and its founder before I go in, but I am still not sure what to expect. We start walking around, and it is beautiful, but it takes a moment for me to understand what I am seeing. Finally, walking down a narrow, winding path through tall herbaceous planting, hidden glories around every secret turn, my mind connects it all, and for a moment my breath stops.

It is a stunning garden in the English countryside. There is a pond and beautiful pondside planting, precise topiary punctuating layered, lush and bountiful borders, expertly combined. But the combinations of plants are entirely unexpected. Side by side with the salvias and grasses and dahlias and common stalwarts of any late summer, beautiful English countryside garden are planted a whole range of exotic, tropical-looking plants, alien to this part of the world. Common garden astrantias and thalictrum weave between statuesque beauties that I do not know, but feel immediate affinity for – our DNA clearly written in warmer climes. Unlike other gardens I have seen, here these unusual plants are not othered, not set aside as trophy display items to be ogled in their strange juxtaposition to the plants that make up the norm. They are not caged in glasshouses. They just live here; larger, different leaf shapes and vibrant colours all mixed up with the other plants, feeling completely at home. They look entirely like they belong. At first I think that they could not possibly be hardy, but on leafing through the catalogue of nursery plants it turns out that they are, most of them left in the ground to die back and return like any herbaceous perennial. And how they thrive! Not in some subtropical south-coastal garden, or sheltered in a hot London microclimate like so many of my kin, but right here in the middle of the Somerset landscape, in a garden not far from my own.

As I walk along the paths, hand in hand with Oli, a tropical transplant intertwined with a native son of the soil, I recognise the garden I hope to create. The one that tells our strange and exotic but ordinarily English story of uprooting and migration, of comings and goings, of Imperial legacy. One that composts the legacy of pain and uses the rich soil of honest history to grow something beautiful and new. I feel how I want to grow alongside these plants, claiming my place on foreign soil. Like them, I can be fully hardy here, in this countryside. Oli and I stand above the garden's deep pool and watch our wavering image, broken by water lilies. This special garden has shown me to myself and in its reflection I am inspired.

Harvest

Mexican Hydrangea

WE ARE LEAVING the garden.

It is the very end of the school summer holidays. With the lull in rates of viral infection and morbidity, the hospital has lifted its restrictions on staff holidays, and Oli has been able to squeeze in a much-needed week off before both children start school. We make the decision to brave leaving our new home for the first time since moving there and drive up north to Cumbria, to the small shepherd's cottage that Oli's grandmother bought many decades ago as a family holiday home. Sitting up a narrow track in a field clinging to the side of a hill, with running water and electricity recent arrivals, and a perpetual case of rising damp, it is hardly a luxury escape; but stepping from the confines of five hours in the car into the enormity of this landscape feels completely blissful. His aunt now lives there permanently, but Oli, the children and I stay in the small room that has been made habitable above the adjoining shed, all four of us sleeping together for warmth in a house with floor tiles that lie directly on uneven mud foundations, and no central heating. Bracingly fresh air will be abundant; the children relish the adventure.

The house is humble, but the views are glorious. High

in the hills above the Eden Valley, sitting on the bench along the stone front of the cottage, buffeted by the wind, the horizon stretches out for miles. The landscape is rugged here, not the soft, embracing folds of land that we now live among in the southwest. Jagged cliff faces claw their way to the sky, fells scar the land, bruised purple with heather and thistle. What few trees punctuate the steep hillside are near-horizontal, bent over by the prevailing winds. The children run out into the field chasing rabbits; in the face of the full force of the wind they open their arms, cast their weight onto the air whipping around them and try to fly.

From the beginning of our relationship, Oli has brought me to this cottage perched more out of habit than any structural integrity on the Cumbrian hills. Here, he taught me to brave being in the British outdoors. It was on these trips that I bought my first walking boots, acquired waterproofs and wellies. It was while trying to make my way across the landscape of Eden that I first unfurled an ordnance survey map. It was in the clear, peat-stained waters of High Force that I first swam, shrieking with cold, then endorphin exuberant, in English waters. Over a decade and a half he patiently guided me through this landscape, but also wordlessly tutored me in the vernacular of the culture of the British countryside. And walking beside him, a White man whose belonging was unquestioned, eased my way in. When I walk alone on the interconnecting paths that trace a fine web over the landscape around our village, it is all those years of induction in the language of country living that mean that in this still-new rural setting I do not feel entirely out of place.

Our week in the cottage in the hills is the restorative break we all need. We feel freer roaming across the expansive landscape. The harsh winds strip away artifice, the miles of horizon seem to leave me as open as the wide-ranging views. I see it in other people too; the locals feel different here, more open and direct without the opaque politeness of the south. For all that I stand out like a brown thumb, I feel welcome in this vast vista that has room to accommodate so many. It was in this revealing landscape that I first fell in love with England. I love it as I love the man who introduced it to me.

In this huge, impassive landscape I feel insignificant. I lose myself from the centre of all my petty concerns, and in that loosening find myself reconnected, a part of a bigger picture of existence once more. Being here changes my state of mind, it provides a necessary reset that we have returned to time and again in the fifteen years or so that I have known this place. And yet this year, with all the upheaval and strain we have been through, I sit huddled with my mug of tea against the whitewashed front of the building and feel a difference in myself as my eyes take in the rapidly changing light over the land below.

We are so high up here that clouds sometimes scud below us, and I watch the play of light and shadow over the fields seamed by stone walls. As the light breaks through higher-level cloud to shine on Wild Boar Fell, it occurs to me that while I relish this holiday that we so needed, this year I lack that sense of desperation that usually accompanies our visits here. I have felt both utterly trapped in our garden by the circumstances, and utterly in love with the intimate relationship that those circumstances forged, but now that

I am out of it I have the space to find that I no longer carry an urgent longing for escape. I probe around in my heart, but that restless urge that I seem to have always carried within me seems gone. The absence feels profound. Despite everything, regardless of my conscious will, some need that I cannot even name has been met within me, through the day-to-day living in our new home, through my time spent in our new landscape. I am reaping what has been sowed. The garden has changed me.

***

The autumn light is a marvel. Last night, encouraged by the clear night sky and the lengthening evenings bringing the return of the stars, I left the curtains open as we went to bed. We had returned from our holiday, refreshed by a change of scene and made more appreciative of this now beautifully familiar one. It was a dark moon, and the stars gleamed brightly in this part of the valley with no street-lights or illumination. I drifted off to sleep looking at the stars wheeling overhead. I sought out the patterns of familiar constellations from my childhood, like the Big Dipper, and gazed, waiting to see if my favourite would appear: Orion. A trio of moles mark the skin of my forearm and I had always imagined them as an earthly echo of the starry figure's belt. I fell asleep before I could spot it.

I wake with the light in the morning and bury myself more deeply against the returning chill. It is the weekend, so I stay in bed a moment to luxuriate in the view. The sun has not yet risen, but the lightening sky is pale blue over the trees, a sliver of field and the summer growth of our hedge. Sheep have been returned to the field to make the most of the flush of regrowth that followed the hay cut earlier in the summer. I idly watch them grazing, with the waking crows flying back and forth from their colony in the wood overhead. A movement at one corner catches my eye; a fox slinks through a gap in the fence back into the wood. I think I can see something pale brown in its mouth and wonder if my neighbour has lost another chicken.

I stretch and sigh, eyes still on the view framed by the pair of wooden sash windows. This is the sort of vista that only a few months ago I would have had to save up and wait for on holiday; an escape, a respite from the brick walls and rooftops, from the streetlights and traffic and wailing ambulance sirens. But here I am, in my own bed, sleepily gazing at the sky, and trees, and animals, and birds, the only noises in the early morning being the calls of the various creatures beginning their days. Everything about the scene infuses my being with a sense of peace, and an ease that was so elusive in our old life. Here I am steeped in it, and have only to pause and allow myself to soak it in, yet that can be so challenging.

Eventually, I get out of bed and go downstairs to wake the house. As I pull open the curtains over the French doors in the living room, golden rays stream through the woods as the sun rises behind the trees. The dappled light dances over the sofa, filtered through the wisteria growing over the balustrade of the deck and partly covering the windows. The spiders have been busy as autumn approaches, and their webs between the metal railings that edge the deck sparkle with dew in the early light. There is something magical about this light as we approach the autumn equinox. I stand in front of the doors, feel the sun's warmth despite the lingering nip in the air. I close my eyes, and the sun's rays on my face summon an image from the depths of memory.

I am lying on the beach, bamboo woven mat against the skin of my back, grainy sand between my toes. My eyes are shut as I drowse in the caressing warmth of the sun, the smell of salt, and sea, and warm sand enveloping me. The whoosh and hiss of the waves pounding the beach regular as a heartbeat.

There is a photo which captures me in one of these moments, probably taken on one of our Easter trips to Blanchisseuse, a more distant beach on the north coast of the island than our regular weekend jaunts to the popular Maracas Bay. In the ageing photo, my teenage self lies stretched out on the sand, one arm slung over my head, eyes closed, face upturned to the sun like the flower I am named for. My parents gave me the middle name Ayana: beautiful flower. I stand in my living room basking in the autumn sunshine more than two decades later, a sun-loving flower still, despite all that has changed in the intervening years.

So much has changed. I watch the light climb and build over the trees opposite, the very first reddening leaves at the tips of the branches of the field maple glowing like sparks. The edges of the tree look like embers in the morning light, radiating life-giving warmth. A gentle, sustainable heat.

'You need cooling.' My grandmother's words come to mind. The burble of the stream at the bottom of the garden catches my ear, and the water washes my thinking along with it. I have had enough fire; it has driven me for so long. What I have needed is a wellspring, a source of a cool perpetual flow of life-giving substance, like this stream. To raise the level of the groundwater. To sate my drought and fill my cup. The garden has been my source all this strange, difficult spring and summer. Landing in this damp, cool garden was no random impulse. I recognised the environment that I needed to root myself into.

The focus of my gaze changes as I notice what I first think are motes of dust blowing around, caught by the light. When I look at them more closely, I realise that they are tiny insects dancing above the sodden deck and greening earth below. When I first started my psychiatry rotations, driving between different hospitals on shifts, my windscreen would be filthy with insect bodies in summer. By the end of our time in Oxford the car was always clean. In twenty years I lived through the equivalent of an insect genocide. I have grieved and felt anxious for it; but in the way that our brains habituate to shifting baselines, I had become inured. Now I stare in hopeful wonder at the cloud of tiny creatures in front of me.

When the ground source heating works were finally completed after all the pandemic-induced delays, I found it hard to imagine that the scarred site of such upheaval and destruction could ever be beautiful again. And yet over the summer a dense green flush has quickly grown up to cover parts of this earth, so unlike the compacted beds that sat empty for so long this spring in the terraces above. There are still many bare patches, but between them are wildflowers, come of their own accord. Seeds blown in from the fields around, or awoken from beneath the neat lawn under which they had lain dormant for years. With the plants have come this other life, tiny creatures dancing and swirling above the land below.

My throat catches as I think what lush growth could come to settle here from the upheaval and destruction of our old lives, what joyous ease we could find through learning to dance rather than fight our way through our days. So much has changed, but so much can change still. This cloud of insects claiming their home so quickly on its return contrasts sharply with my recent realisation that while our lives were completely altered months ago, the change in my mind has lagged behind.

I have been extremely grateful for riding the storm of this year out in the garden's harbour, and wished that sanctuary was available to all who longed for and needed it. I have fallen in love with the garden, and felt my heart fill to bursting with joy and wonder for the space, felt it absorb and transmute my rage, felt its gentle comfort in my deepest grief. Yet somehow I have not been able to fully believe what I have had. As I stand here before the doors of the deck, a fugitive in this liminal space between earth and sky, I recognise that a small but crucial part of myself has been holding back, and acknowledge the old, under-lying fear that has been getting in the way. I am afraid that I will be rejected from this place, told that it is not really for me. I am afraid that if I let myself fully love this home, old patterns of abandonment will be re-enacted. Somehow, I will be uprooted, and torn away from this beautiful patch of earth. In that leaving, the garden itself will abandon me.

There is a reason why there are not many like me in villages like this. Why, for so many Black people, it feels safer to band together within cities, why the countryside is viewed as no haven. The fear is logical and understandable. Even when the internal severing from the land that is the legacy of slavery is surmounted, it makes sense in response to the ongoing hostile depiction of the countryside as an unsullied, 'English' space. These constructions that we project onto the living landscape around us are unnatural, but racism has imbued every aspect of the human landscape that we travel through. Every defence mechanism that I have employed has been necessary for my survival.

And yet here, in this softly receptive plot of earth, in the welcoming community that it grows, that has challenged my preconceptions of belonging, I can feel how fighting for survival is killing me. This year I have been trying to grow into a different shape, one that suits the soft contours of the ground around me rather than the harsh Imperial terrain that has shaped my mind, but the rigid armour of a lifetime has constrained me. The land demands something different of me – a vulnerability – for me to truly begin to get to know it. This abundant setting calls for softness, and ease, and joy. It evokes thriving. And yet, because of all the years spent longing for those things but not expecting them, having to work hard to stave off their opposites, it does not come easily to me to yield to them. They have been denied to me – to so many – and to those before us for so long. They have been written out of my body.

In this garden a different narrative to the one I have been taught grows. A rich story of collective abundance, of cycles of light and dark, life and death, growth and rest. It holds me, but I have lain stiffly in its embrace, not trusting enough to wholly mould to it. I have tried but not yet fully yielded to its loving invitation to shed my defences, bare myself and soften into its contours. It will take time – it means laying down the layered burden of many generations. The garden has changed me; it has grown into that restless void that I had always known. Grown and died, and rotted down, and grown and died again so that rich fertile soil now fills the heart of me. But there is still so much change to come, before the garden lives so deeply rooted in me that it will spring from the earth around my feet wherever I make my home; before the garden that grows within me is so lush and abundant with love that I need not fear abandoning myself.

I look at the seeds that have germinated in the soil beneath me. They split their coats, buried emerging radicles in the ground beneath, thrust tiny shoots to the sky. I think about what it would feel like to split my skin and root down into the earth, dissolve my current being and transform into one capable of shooting upwards to harness the power of the very sun. I am awed and humbled: this must hurt. And I understand the pain of this past summer. This is what change feels like.

As I stand here watching over the insects and the wood, a movement from the just-reddening maple catches my eye. A leaf is blown off in a gust of wind. Soon the entire tree will be bare. The world around me is shedding itself before the earth, and softening into the bare tenderness of autumn.

***

Returning through the garden after the rush of getting the children off to school on time one morning, I linger outside. It is unusually warm and sunny, so before getting on with the housework and other tasks I need to do, I let myself be drawn by the playful golden light. I climb the steps of the terraces, heading for a brightly lit spot near the top of the garden, from which I can get an over-view of the site below. A patch of flowers shines like the sun in the shade beneath the steps I am aiming for. A group of rudbeckia has come into profuse flower just next to the path. A clump of tall, white daisies lies behind them, and in this now-part-shaded bit of the garden the daisies fall clumsily over themselves, stumbling into the path. It seems clear to me that I should try to move them to a drier, sunnier spot for them to really thrive. But the rudbeckia surprises me. It was a present from my in-laws' garden, planted here in early spring before I discovered how shady these beds would become. Another sun-loving plant, it seems unexpectedly happy here, standing tall, glowing faces raised to the sun with their black button eyes. The garden lets me make no assumptions.

As I brush past the rudbeckia before climbing the steps, I see it. A new bloom that I have not noticed before. The plant was extremely late into leaf, standing as a set of seemingly dead brown sticks near the back corner of the bed, not far from the still-glowing spiky heart of the terrestrial bromeliad. Opposite pairs of large, velvety, heart-shaped leaves eventually emerged, flushed deep pink at first, then with purple veins on dark green, which attracted my interest. They had then sat seemingly unchanged for the last few weeks, until this.

At the top of one of the stems has opened a bright pink puff of colour. As I clamber through the bed to get nearer, I realise that each flowerhead holds a large cluster of tiny, vivid pink flowers. The overall effect is so similar to the ixora flowers of my grandmother's garden that I reach out and pluck one, almost able to taste the sweet nectar that each flower held. As a child I would pick the tiny ixora flowers one by one, to suck the drop of honey-sweet fluid at their base. On closer inspection I realise that, for all their mind-blowing similarity en masse, these flowers that have appeared in my English garden are different. Each individual flower has five petals, rather than the four I remember, with a cluster of pink stamens on thin filaments emerging from its centre, rather than the single central thread of the ixora, which could sometimes be pricked out, carrying its drop of nectar, with delicate fingers. These flower petals join together at their base to form the same tube as the ixora. They smell sweet, but there is no nectar to be sipped through its narrow straw. But the memories that the bright pink flowerheads unleash of the first garden of my childhood – of my den deep in the ixora hedge, where I would play and make secret magical potions for the fairies I was certain lived there – are as sweet as the sips of all those decades ago.

I am giddy to see something that reminds me so much of my childhood in the garden. A plant so nearly the same, but different, its own creature, a new beauty hardy to this place. I look it up on my phone and learn that it is called the Mexican Hydrangea or *Clerodendrum bungei*. The more I read up about it, the more amazed I am that it is here and flowering in the first place. It is hardy to only minus five, and likes full sun and well-drained soil. Here it is in a frost pocket, in a quite damp and shady part of the garden. And yet it is flamboyantly thriving, a burst of vivid colour against the slightly fading background of my early autumn garden. I love the transition into autumn almost as much as I love the burst of spring, but the sadness of the winter to come always colours this time of year, and adds a mournful edge to its beauty. The vibrant joy of this flower counteracts that spectacularly. I gaze at these bright pink, not-ixora flowers that have appeared at just the right moment, and feel so grateful to this space that keeps providing me with what I need.

***

I stand in front of the doors to the deck, as I have done many days this past summer. A pair of pigeons made a nest in a particularly inconvenient position in the wisteria right above the door. After wildly startling the mother bird a few times, whose alarmed presence curbed any relaxation from sitting in the space, I gave the deck over to her family for the last few weeks. Every evening the pair of adult birds flew down from the nest and performed a ritual on the balcony railings. It looked like a dance of greeting and reconnection, ending with them both grooming and nuzzling each other, before one flew back up to resume the vigil on the eggs. But summer is over, and this morning things seem to have changed. The atmosphere out there is lighter. I open the door to the deck and no panicked flapping greets me. The baby birds must have hatched and flown the nest. I can reclaim my space.

My babies have flown the nest too. My daughter has joined my son at school, and after a couple weeks of settling in, the entire school days are my own. My life must find a new balance.

Today is the autumn equinox. A moment when all hangs briefly in equilibrium, one of only two points in the year when light and dark are in perfect balance. I find this deeply comforting. The rarity of perfect balance in the celestial spinning of the world round the sun makes the never-ending push-pull of my insignificant life feel more normal, and less exhausting. As well as a brief moment of balance, this equinox is the tipping point of the year from which we will hurtle ever more rapidly into the dark.

I am especially grateful for the space to think that having both children at school has given me this morning, since I feel as though I have lost all equilibrium. One of my mother's cousins has been researching our family tree on the maternal side, and hearing that I am curious about family stories has got in touch with tales of our shared ancestor, my great-grandmother, Emily. Charmaine was a little girl at my great-grandmother's knee, and listened to all her tales. She said that Grandmother Emily was always telling stories, and told her that she must remember them, and pass them on, because someone else will need to know them one day. I cannot shake the feeling that my great-grandmother somehow meant me.

Charmaine emails me the story of Emily's birth on the island of St Vincent, how she moved to Trinidad after finishing school in search of work. She met Robert, my great-grandfather, who had come to Trinidad from Barbados, and got married. They had seven girls, of whom my grandmother was the youngest. We are into a time that I know now, a family history with which I am familiar. But she tells me more, reaching back into a place where all the stories that I have are shadow. Her words add substance.

Emily was the daughter of Jane, who was born by a different name to one of the Indigenous tribes living on St Vincent. Jane's parents were converted by Christian missionaries, and her birth name lost. Jane met and married a British missionary who had come to the island with his sisters to save pagan souls. Jane and Mr Sunderland had three children, one boy and two girls, one my great-grandmother. Charmaine tells me how Emily told her that her parents married for love. This part of the story was important to her: a Carib woman and a British man met and married for love.

My entire world tilts on its axis. I have always known that Black and White bodies came together in communion to make mine, through the various violences of colonial Empire. But I always imagined those unions to be unholy; non-consensual clashes of spirit, body and mind, raping and pillaging my bloodline. But here was my Indigenous great-great-grandmother at last. My grandmother's grandmother. The one I had always imagined, dreamt of, wondered about. Here she was in story made flesh, emerged from the shadow of whispered speculation and self-denial. Here she was, despite the painful erasure of the version of herself that I am desperate to know, held softly and tenderly in relationship with a native son of this United Kingdom's soil. I had heard fragmented stories of other pairings of great-grandparents in relationships willingly entered – an ex-enslaved great-great-grandmother happily marrying an Irish man – but they seemed so unlikely, and were told at such remove that I had doubted them. Dismissed them as mere wish-fulfilment to soothe the painful truths of enslavement and plantation life. But here was the directly remembered story by Jane's daughter of her childhood. In its loving retelling all my narratives are rewritten.

My mind is reeling. The perception of my self, of how I have come to be, of the story of my existence, has been uprooted in a moment.

Jane. I mourn that I do not know the name she was born with, as the last Indigenous woman in my maternal line, whose mitochondria my cells carry. The last of my ancestors born into right relationship with the earth, before Christianity colonised her. The story is complex; not all facets to be celebrated. At the same time I am exhilarated to learn that there was more nuance, more layers, more love in my recent ancestry than I had dreamed possible. The poisonous trauma and violence that has saturated the soil of my heritage has been pierced by a deep root of love. Love is my birthright, my belonging. It feels liberating.

***

Over breakfast I sit in the conservatory and watch the morning light wash over the garden as the sun rises, a spreading benediction as the minutes pass. It is a beautiful day to garden.

Oli plans to take the children to market with him, and I look out, making a list in my head of the jobs that need to be done in the couple of hours that I will have to myself. I can see that the shrubby germander's silvery foliage has grown far out into the garden path and needs pruning back. Some of the plants that have not done as well as I would have liked in the positions in which I tried them need moving. Others have thrived so much that they need lifting and dividing. I look at the pattern of the light over the garden and mentally check off which perennial plants might be big enough to divide up to fill some of the gaps that remain. I have never done any of this before, but I feel more sure than ever in my instincts about what the garden needs. The thought of having a go makes me feel like a real gardener. I linger over breakfast, thinking about the garden work, about tasks that will make me tire and sweat. But it is so enjoyable to plan my time in the garden in this way. It is a form of play.

Play is the core developmental process through which we create: our selves, our relationships, our worlds, our meaning within it. It is vital for our wellbeing, and yet it is an area we tend to not only neglect, but completely denigrate, as adults. We see play as silly, or lazy, getting in the way of the serious work to be done. Yet play is how we create anything new. It is how – by letting ourselves enter that transitional in-between place of limit-less potential – we solve dilemmas, and find our potency to innovate solutions to problems. Without play, we lack the mental capacity to imagine the creative new worlds we all need now. Instead we stay trapped in old ideas, chained to our past, enslaved to our extinction.

I am playing with the garden. I am trying to hold it lightly so that taking pleasure from the space does not become a form of domination. I will be no conqueror here. I have had so many serious moments with the garden this year, but I want my relationship with this space to be multi-dimensional, for the garden to be able to play along with me. I want to try to listen to its responses to my moves, so that together we can create something new. It has occurred to me more and more over the months of being in this space that gardening in this way feels akin to being a therapist.

When I sit in a room with a patient, I listen carefully to their words, noticing the things they say, and often most especially the things they do not. But communication goes well beyond that. Only a tiny fraction of our communication is through the words we speak; the vast majority is unspoken via the rest of our being. And so it is that I try to listen with my entire body in the therapy room. To the flickers of expression that dance unthinking across the face, to the changes in vocal tone, but also the sudden clenches in my gut, or my jaw, as the emotional atmosphere in the room shifts. I am listening with my unconscious mind as much as my conscious one. And holding the space between us lightly, an internal place of growing potential, where the give and take of our thoughts becomes a form of mindful play. Through the deep relationship that is built, something new is co-created between the therapist and the patient.

This communication, this way of being in playful relationship without words, might underlie much of our capacity to relate to the more-than-human world. It seems instinctual, and feels the same as the deep listening of early motherhood. It is the same as forming a relationship with the other-than-human beings of my animals, or houseplants.

So while I cannot fully understand it yet, it seems to me that the garden speaks as we do. The absence of human words hardly matters when they form such a small part of what is said. I have so much to learn still of the garden's language, but I have the tantalising sense of something that is just on the tip of my tongue. A name or a phrase whispered on a memory, not quite audible. Something that lurks there just out of reach of my conscious mind, that might come back to me in a moment.

So I immerse myself in the garden, and playfully pay attention with my whole being. I try to listen to the space rather than only tick off a list of jobs that have been dictated to me. And as my understanding grows, so does our co-creation.

***

Slicing through the overgrowth is comforting. These geraniums behind the stream, at the foot of the ornamental blackcurrant, have grown hugely lush and wild. Geraniums are not a plant I have thought much of before, my eyes glossing over the neat mounds of ruffled leaves, backdrop to their simple pink or purple flowers, filling space at the fronts of tidy borders. They are the sort of plant I would have classed as useful but boring in my growing internal lexicon of English garden plants, had anyone asked me.

This summer, in my garden, I have fallen in love with them. There is one in an awkward spot at the foot of the beech hedge, a shady place where the soil seems depleted and dry, which flowered prolifically earlier this year. It formed one of those small, neat mounds, with very prettily frilled leaves, but the flowers took my breath away. They were pink, but what a pink! Blushing petals with veins marked in a fuchsia so vibrant it was shocking. At the centre a cluster of pale yellow stamens surrounding an almost coral stigma, each flower a magnificent sunset in exquisite miniature.

Then there is the pair which emerged from the scrub revealed under the bamboo after it was hacked back. These were slightly bigger plants, with more standard leaves on longer stems that flopped about a bit as the summer progressed, but again, the flowers won my heart. A fairly standard geranium purple-pink, but the thing that drew my eye, which I could not capture with the camera lens on my phone no matter how I tried, was the glistening iridescence they carried. They shimmered like jewels whenever caught by the sun. As did the blooms on these clumps with their darker blue flowers at the back of the main ornamental bed that I am currently hacking through. The geraniums are mounds of glimmering magic sprinkled throughout the garden. Who knew? Other gardeners, clearly. I understand their ubiquity.

These geraniums did not grow like the others. Their leaves were bigger – when they first emerged a plant identification app told me they were aconites, and triggered a small panic about inadvertently poisoning the children – and they have sprawled all over the surrounding plants as the months have gone on. They're all leggy and jointed, climbing up the blackcurrant, which has produced one small bunch of fruit, mingling with the nearby huge clump of pheasant grass, which really needs dividing, and seemingly smothering my favourite iris. This was what prompted me to this task of cutting back. It seems an odd thing, all the reining in that the various horticultural missives that I have signed up to, follow online or watch on television are constantly reminding me to do. It is advice I have largely ignored. I chafe against the feeling of tight control.

But I understand the urge. It was Lughnasadh, or Lammas, at the start of last month, an old Celtic festival of the first harvest. It seemed to mark a tipping point in the garden, when its growth expanded past any hope of my control and exploded into its own, full, being. Feeling it happen, and watching the accelerating expansion that continues as we head toward the autumn equinox, with the sense of complete loss of control that accompanies it, has been both glorious and terrifying. It is a feeling of letting go, one that I both desire hugely, and also fear. It is a sense of freedom.

In touch with my ambivalence about freedom, here I am curtailing this geranium's. I gather up the piles of long stems that I have cut off with my sickle and stuff them into the bucket next to me to take back to the garden waste bin. A compost heap remains on my wish list. I look at the neat mound of leaves that I have left behind at the plant's centre. Apparently it might produce another flush of flowers this autumn after this treatment. I do not have a sense of the plant seeming unhappy after my taking my knife to it. It is an ornamental garden geranium after all, probably many steps removed from its wild cousin, developed in co-evolution with gardeners to be regularly preened and primped in this way. It has been selected to perform for human appreciation in this singular fashion; it probably does not mind the cut. A bit of me still feels sad for making it, though. For taming its long, silly, draping, joyous attempts to clamber over its neighbours and reach for the sun.

But mostly, I feel better. After a lockdown spring of spending many hours every day in touch with the soil – mulching, sowing seeds, planting out, and a sort of general going round touching and looking at things in an often non-specific way that my mind lumps together under the idea of tending, but which seems to create the relationship of knowing each other that has developed between the garden and me – the summer was spent at a more distant remove. Forced indoors by hay fever, and then pinned by the inertia of languorous high summer, and distracted by a return of busyness in life as the near-universal pause of earlier this year gives way, I came to feel a bit out of touch with the garden. Still connected to it by all the time spent eating meals outdoors, picking flowers and walking around appreciating them, making my garden teas, and watering the veg. But I had not done much by way of getting my hands dirty, or touching the soil.

It feels good to be back, kneeling here in the beds, answering this season's summons. As I stand and survey the results of today's work in the garden – space around the geranium cleared, the large clump of irises that it had covered now split and replanted into three, and a few new plants settled in the gaps – I feel a sense of relief that is not mine alone. The garden is pleased to have me back.

Asters

HARVEST SEASON IS here. Despite our amateur efforts, we are getting a decent yield from the veg beds, and every trug of produce that I carry down to the kitchen astonishes me. There are beans now, long, flat runner beans, and deep purple French ones, so beautiful on the vines that I am reluctant to harvest them. We get a few half-formed sweetcorn, and resolve not to bother again as our garden simply is not sunny or warm enough. We are spared the fabled glut of courgettes, and I vow to try them again next year, this time mulching with comfrey and being more regular with my watering. Slugs have definitely got the lettuce, and the carrots are ill-formed, but we have deep purple beets, more spinach and chard than we can eat, and watercress grows prolifically in the stream of its own accord, so we have not wanted for salad leaves. And now there are a few hard-won squash, and I watch the developing fruit carefully every day as it is Oli's favourite.

I realise that to grow things for those I love has been the greatest pleasure of the garden. I have picked dozens of bouquets, and given them in thanks, for birthday presents, left them as socially distanced pick-me-ups on doorsteps. And now that the food it has pained me to grow is coming into harvest, I am learning that the profoundest joy comes from nurturing the ones I love with food that my hands have helped to produce from this earth.

I pick leaves of spinach, the seeds of which I sprinkled onto compost in the spring, watered and thinned, fed with nettle tea brewed from the field, fretted when they were attacked by slugs, cheered when they survived and grew large and strong enough to be unharmed by a bit of slug damage. The food is so much more than just food for our body; the cycle of relationship that it has helped to grow between the garden and me sustains my soul, nurtures my relationships with all who benefit from my garden's abundance.

I pick the first squash. I take it back to the kitchen with a collection of herbs and the spinach leaves. With them, I cook a long, slow meal, redolent of home. Beans and spinach flavoured with herbs, a choka of squash and tomatoes, a cut of lamb reared by a collective of neigh-bours in a field in the village centre, short hop to stew in my pot. I send a picture to my parents, that we might virtually share in it together. I miss eating with them. We lick our fingers; full bellies comfort heartache.

Later that evening, Oli and I walk through the garden looking at the lawn behind the house and discussing our plans. He has been suggesting that we build new veg beds, to increase our growing space, and to make use of some of this flatter ground nearer the house. I understand why I have been unenthusiastic about his ideas, and wanting to keep any veg-growing – and the unconscious pain it carries – tucked away in a back part of the garden where I do not have to face it.

But I feel deeply now that we have an incredible opportunity. I have no home but where I make one, where the earth herself softly yields beneath my feet. I have changed only the planting in the garden till this point; this would be our first act of making a mark on the landscape itself.

We can build something new, change the shape of the garden to make a beautiful space where I can work through my grief to grow in it. We can have our children come to know and to love the relationship with the land that feeds us. I have a chance to reach back in time to before the trauma that pollutes my recent history. To the time before my ancestors were enslaved, and their loving ties to the land were torn apart and replaced with bloodied chains. I have a chance to take hold of the love my great-great-grandmother grasped, and claim it as my birthright. In this garden I have the freedom to play, the space to lovingly create something new.

I turn to him, and surprise him with my tearful enthusiasm. Yes, let's create a kitchen garden down here where we, and everyone who comes to our home, can see it. Let us build something different and new. Together let's grow something beautiful.

***

I circle my autumn garden thinking about cycles of hurt and trauma caused by Imperialism. Walking round the plants dying as we near the end of the year, colonialism is dying all around me. The path through an English garden is a journey around the globe. Generations of plant hunters brought them here from other lands, just as they changed the landscape of my Caribbean home. Today's English gardens are colonial products, plants moving with people along the field lines of Empire. Today's English landscape is a colonial product, stripped of its own indigenous creatures by the industrial land management practices that colonialism bred. Humans created the changing soil and climate conditions that allowed the plants brought by humans to thrive and become invasive. Every soil has been poisoned by colonial greed.

I pause in front of one of the large pots we brought with us to the garden. We have not planted it out, not yet decided on the ideal spot – it feels important to get it right. The plant is in the final flush of flower, one last burst of heavily scented bloom before winter's death.

It is a rose. The English rose. The beloved symbol of the idyllic English garden: a stone cottage-front covered in the tumbling charm of a climbing rose, or the formal garden of a majestic country house, roses tied up in box hedging knots. This is the plant that most captures the English imagination, red roses against white in a centuries-past war. A pretty girl, an English rose, the only type of beautiful flower that I, Ayana, can never be. My grandmother's beloved flower. A representation of how thorny it can be to love.

And its cousin, the apple, equally claimed in the illusion of native belonging. Gleaming rosy red on gnarled old branches in a picturesque orchard meadow, redolent of the Somerset valleys in which I now stand, the very slopes of my garden holding the ghosts of orchards past. And from my past: foreign apples piled high at the green-grocer's every Christmas as a special English treat.

And yet, for all the potent symbolism of the rose and England, it is not native here. It is an immigrant who has been loved, and naturalised. Roses were originally cultivated for gardens thousands of years ago, probably first in China. Of course, plants have always come, and people too. Both my home islands sites of migration and exchange for millennia. The denial of this is the basis of the delusion of the mythic rural idyll, static in collective imagination, our minds trapped in this common lie, false nostalgic imagery projected onto land we can no longer freely roam.

The Enclosures Act destroyed rural life. It drove peasants living off the land, with their skills in subsistence and their status in village communities, into helpless poverty in cities. There the newly landless and disenfranchised formed a huge labour force with no rights or power. The Industrial Revolution was built off their backs. The same Industrial Revolution being funded by the Slave Trade, and the plantations grown on the claimed and enclosed land in the Caribbean. My ancestors there and the ones here, natives all, seemingly divided and controlled by the invention of racism, were really in solidarity all along through Imperial bondage.

My childhood homeland and this new home land are intimately, intricately connected. The hyphae run unbroken under the deeps of the ocean between the islands. Both lands in desperate need of repair. The problems seem too big to fathom a solution. All I can do is start with the earth beneath my hands, fumble my way through the soil towards loving reparation.

***

I sit in another garden, bundled in layers against the evening's increasing chill. A small group of us have walked here from our homes in the village, the six that are allowed to meet. We are widely gathered around a fire pit, outdoors. The distance between us vibrates with restoring connection.

It has been a strange time for making new friends. I have spent most of the last few months feeling wildly vulnerable, tender and exposed while isolated in the intimacy of a small community. Surprisingly, this has worked in my favour, as none of my usual defences have been able to get in my own way. My heart has been hammered wide open, and into it has been poured friendship.

The excuse for this evening's gathering is our newly formed book club, but the book that most of us have struggled to concentrate enough to read hardly matters as we end up talking about how we have been enduring this surreal time. Our stories are filled with sadness, layered with worry and anxiety, but we are giddy to be together. The alchemy of the space between us begins to transform the tales we share. We howl with laughter, everything suddenly hilarious. It feels almost manic, but also wonderful, for the tears running down my face to be ones of laughter rather than pain.

We all stay far longer than originally intended, and when we finally, reluctantly part, awkward blown kisses in place of the hugs we all need, I head off alone to my garden on the edge of the village. It is one of my favourite things, walking through the village at night. Golden stone turned silver in the darkness, stars wheeling overhead, an air of magic and untold possibility in every shadow.

Stepping through the gate at the top of the garden, I

am greeted by an enchanting sight. A hedgehog, asleep on a pile of leaves on the garden path. It is so deeply in slumber that, walking with my torch off to better immerse myself in the delicious darkness, I nearly trip over it. I bend to look more closely and hear the snuffle of its breath. My heart almost bursts in my chest.

I have known that the garden hosts a hedgehog, hearing it loudly rustling through the undergrowth on recent nights, hoping it has been snacking on the garden's many slugs, but a sighting has proven elusive. But here it is, seemingly so safe, so at home in this space, that it has felt able to just curl up out in the open and snore without concern. I remember how wary I felt of my welcome in this place when we first moved in, and how warmly accepted I felt round the fire pit tonight. I leave the sleeping creature and make my way down the garden, feeling more eerie in the slumbering dark, wondering who else is sleeping in the space with me. Back in the house, I creep into the children's bedroom to listen to their noisy breathing in the dark. Somehow it seems that a home that is safe for the hedgehog will be safe for them, and I am grateful for this place that holds us all so lovingly.

***

At October half-term we decide to make good use of our time at home and build our new veg beds on the lawn behind the house. Reclaimed oak sleepers are delivered by a local yard, and my husband sets to with his power tools, the children and I recruited to help lift and hold the beams in place. Six beds, with the grass left as paths in between; sterile lawn converted into fertile growing space. For as long as the old, steeply precarious veg beds hang on, we have doubled our dedicated food-growing space. Inside each raised bed we lay down all the cardboard we have left from moving house, and empty barrows and barrows of compost on top, dancing on it in celebration of their completion afterwards to knock out some of the excess air.

Straightaway we plant the beds up, sowing some with a fast-germinating winter cover crop to help the soil mature for growing vegetables next spring. Others we fill with winter onions and shallots, garlic and huge cloves of the leek relative called elephant garlic. I bury my hand to my wrist in the friable soil, submerging each bulb in the cooling earth. They will not be harvested until early next summer, and there is something about the act of plunging these bulbs in, in hope of their growing over the months of winter, that feels like a profound act of faith.

Coming from the constant warmth of the Caribbean, I have been stunned by the realisation that gardening has given me of the true brevity of the growing season here, at least for the annual plants that make up most of the food production we tend to depend on now. Some days I wonder if it was simply that which drove the greed of Empire: a search for better weather and more growing time as it became dependent on a few crucial crops. Without the protection of a greenhouse or polytunnel we can grow here for only about six months of the year, unlike the year-round growth that is possible in my tropical home, provided that one has stored enough water in the rains to last the season of drought.

I enjoyed the seed-sowing of last spring and found it therapeutic in lockdown, but given my laissez-faire approach to the houseplants, growing lots of plants from seed every year seems a more intensive practice than I can sustain. Instead, I have bought books on foraging, and perennial edibles, curious about less time- and resource-intensive ways of finding and growing food. Eating more perennial plants would vastly extend the season and narrow the hungry gap. I think that the Indigenous people of both my islands would surely have known about them and relied on them more. There are a few that I find particularly intriguing and plan to order for next spring: a type of perennial kale that seems really quite attractive, near-neon tubers of oca, and amusing but hopefully useful walking onion. I have been reading about a type of largely perennial potager garden that seems like it would suit this woodland verge that the garden seems to want to be. It is called a forest garden. Standing up to stretch my legs from burying next year's bulbs, I squint and look at the valley before me, trying to imagine it more densely wooded, more wild, more populated by creatures, even more beautiful. I look at the veg beds and it strikes me that thinking about planting perennial edibles feels like a permanent move towards growing a nourishing home.

The children run shrieking with laughter, balancing on the edges of the newly built beds, immediately creating a new playground. They ask us if they can live here for ever – a profound shift from months of asking when we will return to our old city home.

***

The garden is melting into itself. Plants shedding leaves and stems, disintegrating into the damp earth now that the autumn rains have come. Everything is ablaze with fiery autumn colours, but constantly wet. Discarded acer leaves gleam like rubies, the coral bark maple drips yellow diamonds. In my imagination autumn always sounds like the crunch of leaves underfoot, but here it is a spatter of rain on stone paths and the squelching of patios covered in a litter of leaves which never fully dry out. Water seems to drip from weigela branches arching over the path whether or not it has rained that day. The long petioles of the tree peony's huge leaves turn fuchsia pink and glisten wetly.

Most of the garden is soggily past its best. Most of it, but not all – it has been unusually warm, and a few plants have flowered unseasonably. A lone foxglove rises in one of the beds, its pastel pink disorientating among the autumn leaves. I measure my life in leaves now, and spend what seems like most of my time sweeping the paths clear. Also my box of bulbs has arrived, and in the beds I brush mounds of leaves aside to find gaps where I can bury next year's treasures. I feel an excited thrill as I move through the beds plunging tulips, fritillaries, muscari and alliums into the earth. A promise of joy for next spring. I plant crocuses and daffodils under the ash tree at the top of the garden. The difference in the soil in the areas that I heavily mulched last spring is astonishing, and delightful. I crumble the dark earth between my fingers, remembering how sticky and heavy it felt just a few short months ago. It is amazing how much can change in a growing season.

I read in a conventional horticultural guide that I should

gather up fallen leaves into bags and set them aside to convert into leaf mould, which I should then spread as mulch on the same beds from which I gathered them at a later stage. This seems like unnecessary work, redundantly doing what the garden will surely do better for itself in situ, so I resist the urge to sweep the entire garden clean of the ever-falling leaves, as if it is my living room, and stick to the paths. I settle for brushing leaves off the crowns of some plants that I think will not like a wet winter blanket, and freeing the lavender from being submerged by the tsunami of shedding wisteria. I leave the browning stems of most things standing as this year's visible growth dies back. I read that this can protect more tender crowns from frost damage, and last winter showed how much frost loves this place. I still don't feel as if I know what I am doing, but I have the sense that, at least for these hardy perennial plants, the interventions required from me are for the most part minimal. A light touch of checks and balances. I trace my fingers along the fading leaves and stems of the new perennials I introduced to the garden this year and hope that they will make it through the bleaker months to come.

Reading that many insects overwinter in standing stems and long grass reinforces my determination to leave much of the garden alone. This summer I encountered many insects that I had never seen before; I want to give them a safe home over winter too. So much of the garden often still looks messy to my not-yet-fully-decolonised eyes, but I know that this is how I have been conditioned to regard it, unconsciously imbibing the message that everything must be trimmed back, small tidy crowns of perennials surrounded by large gaps of bare earth supposedly the proper look for a well-cared-for garden at this time of year. I remember how distressingly bare the garden looked this winter, and looking at it now, full of standing stems and seed heads, and a few grasses, it suddenly strikes me as lush, despite the dying back of autumn. The change of perspective makes me love what I see. I care deeply for this garden, and it is becoming apparent that true love might look different to the patterns of expression we have been taught.

Under the carpet of brown leaves, some parts of the garden are still vibrantly alive. The evergreen plants, many of which have sat largely unnoticed by me as my eyes have tended to skim over them in favour of bright summer flowers, begin to come into their own. Beneath the shedding wedding cake cornus, the broad leaves of winter heliotrope shine more brightly. The lime green conifers on either side of the patio look startlingly fresh among the growing brown. Some evergreen ferns that I planted along the stream look suddenly larger and more vibrant as the plants that have overshadowed them all summer begin to fade. I feel grateful to my past self for putting them there, as I see how their green will help mitigate the garden's grey in the darker months to come.

In stark contrast to the fading life all around, one patch just outside the conservatory door is coming into vibrant bloom. All year I have watched a thicket of fat green stems with thin leaves rising from beneath the lilac. Their growth has been dense, and rapidly spreading by rhizomes, colonising the space left by razored bamboo stems. I have felt a little nervous about the voracity of their expansion, and have made some inroads into containing their edges, uprooting clumps that tried to spread into the bed below. But mostly I have been curious to find out what they are. Plant identification apps have been uncertain, suggesting something different every time I have tried, and so I have simply had to wait for them to reveal themselves.

Finally, they come into flower, and I am not disappointed. It is a magnificent clump of asters. Located in the sunniest spot of the garden, one of the few still receiving direct sunlight for much of the day, they bloom profusely, so that it seems the ground is covered in a cloud of lilac stars with yellow centres – a cloud which thrums with bees, the ground almost vibrating on sunny days as seemingly every bee, wasp and hoverfly for miles comes to feast on this late banquet of nectar. I feel glad that something restrained my hand when I was tempted to rein in their spreading shoots. I had noticed the urge in myself, the tinge of fear in response to the invader that would take the place over if we made the mistake of letting them in.

The asters are not the only plant that has grown exuberantly in my garden. A few large clumps of *Acanthus mollis*, repeating groups of alchemilla and thick patches of crocosmia have densely established themselves in some parts of the garden. Elsewhere large stands of Japanese anemone and running thickets of ornamental raspberry have formed. I have started paying attention to the language used to describe these plants' behaviour when I read about them in gardening books and online. Some are deemed 'thugs' and 'invasive', words with aggressive connotations of forcing the natives out. Others attract the more neutral descriptors of 'vigorous' or 'easy-growing'. I cannot quite understand what features make some strongly growing, rapidly spreading plants more tolerable and less feared than others – from my initial amateur attempts to curtail some of them in my garden, they all seem like a pain to keep under control. I do not know why the acanthus is an invasive thug but the Japanese anemone merely one that enjoys spreading through the borders, when they both seem to regenerate just as easily from small bits of root left behind when I try to dig them out. All of the plants are foreigners to this island, imported by plant hunters in colonial times to enhance English gardens with their beauty; but, for reasons that are opaque to me, some are welcome and others are not. These are the same irrationally false divisions that we apply to ourselves. But regardless of what label we apply to them, here they are happily making themselves at home in foreign soil.

The most vibrantly alive parts of my garden as it decays into autumn are the wildest bits. With so much of the space laid out as beds to be cultivated, I have concentrated my efforts this year on the parts we see the most, as we sit in the conservatory or on the patio, or on our walk through the garden every day as we come and go. The parts that were the most barren to begin with. These beds are all much more lush and beautiful, but they are outdone by the less visible bits that I had largely left to their own devices. The garden's so-called weeds have flourished there.

Clumps of nettles grow on either side of where water drains in from the neighbouring field to our stream at the bottom of the garden. The children have watched awestruck as scarlet tiger moths transformed from caterpillars on the plants. The caterpillars shared the plants with me as I made countless pots of tea: delicious ones for myself and a hideous-smelling concoction as fertiliser for the garden. I picked so many nettle tops over the summer that my fingertips are now immune and no longer feel their stings. Next to them is a dense patch of creeping buttercup that put on a brilliant, glowing show in late spring. Cleavers stick to the top boundaries along the hedge; the children have stuck them on my back and run off laughing about sticky willies all summer. Various dead-nettle relatives thrive among the rampant vinca in neglected borders. There is comfrey, which I have fed the plants, and sedge, and bee balm showed us all summer how it got its name. There are wild violets and ground ivy and common speedwell, which entranced me with its minuscule, exquisite blooms. The bottom stream is overtaken with water mint, and angelica has reared elegantly magnificent, almost as tall as the hedge.

I have sometimes felt ashamed of the weedy sections of my garden, of how wildly out of control they reveal the space to be, but mostly I have loved the weeds, have adored seeing them grow lush and large and vibrant, unconstrained in this fertile landscape to which they belong, compared to their warped and stunted growth in the cityscapes where I first met them. I have raided them for tinctures and balms and pots of tea, have delighted in the array of creatures that they host. And found deep joy in the wildness in me that they cultivate, the reminder that I too am part of the 'wild' which has been touched by human hands and trodden by human feet for many tens of thousands of years. The weeds joyfully proclaim that this space can never be fully tamed, and of the freedom available to me if I yield to it. Too often I have been treated like a weed between the cracks. There is joy and a lesson in watching these ones claim their space, widen the cracks of acceptance, and audaciously thrive.

I want to expand what these wild indigenous plants can teach me. We order a native wildflower mix, locally produced and suited for our location and wet clay. We begin preparing the bare patches of earth on the scarred bottom lawn of the garden to receive the seeds in spring. The small cloud of insects floating over the green patches that returned had given me an idea. They gave me hope. I want to see if this garden and I can grow more.

***

There is death all around us in the garden. As the plants decay back into the earth, the bones of the space become more prominent again. While sweeping leaves off the curved stone bench built into the wall next to our new veg beds, I notice that some of the rubble used to build the terraces contains fossils. I run my fingers over the spiralling curves of ammonites and try to understand that the creatures preserved in stone lived more than sixty-six million years ago, with the dinosaurs. From the perspective of my nearly forty years, that expanse of time feels almost mythical. The ammonite could be a fossilised unicorn horn, for all that my mind can make sense of it. Along with the fossils, there are broken bits of stone that must have come from old gravestones. I trace out the partial years from the eighteenth century – much more recent times – and the sections of names visible, and shiver slightly. Old grave- stones and the holy stream; there is much that is sacred in the garden.

Samhain approaches, the sacred Gaelic festival marking the end of harvest and the beginning of winter, seen as a time that is a portal into the world of the dead. In my Catholic childhood, it was the eve of All Saints' Day, a celebration of all who have attained heaven, followed by All Souls' Day, to commemorate those who have not. On these days you visited and tended to the graves of your beloved dead. I think of my grandmother and grandfather locked in the Lapeyrouse Cemetery at home, no one coming to visit their grave in Trinidad's strict ongoing lockdown. Looking at the decay in the garden around me, the ances- tral honouring of Samhain seems to make more sense than the sugar-fuelled celebrations of Halloween. The autumn garden is the year's mourning.

Like the shedding of the autumn leaves, and the decay of summer's growth, mourning is a necessary act in the cycle of life. There can be no new season of growth without the mourning of the one before. The cycle of life speaks to constant change, and the only constant of change is that it involves loss. It is in grieving what is lost that we are able to accept what is gained. Without this process, without the ability to truly mourn, we remain stuck. Lost in detached numbness without even the ability to lament our fate.

Mourning is also a reconciliation, an acceptance, a joining together of split parts into one whole. It is the process by which we mature enough to see and hold both the good and the bad parts of the ones we love. Our love itself becomes complicated, more nuanced, touched by sadness and guilt and grief, as well as delight and bliss and joy. Our love becomes more accepting, less conditional. We become more integrated. We are in touch with the truth of our reality. All parts belong to the one whole.

Mourning composts death into new life.

This year, Samhain falls on the night of a full moon, a blue moon, as there has already been the harvest full moon at the start of this month. The idea of a blue moon feels suitably spooky. To mark the festival, and to mitigate my daughter's disappointment at this year's necessary cancellation of her favourite holiday, we pack a small picnic of cake and a thermos of hot chocolate, and walk to the Long Barrow. The village lies near to a neolithic burial site, and as we walk round the tomb created for humans who lived and died in this spot more than five thousand years ago, I feel the eerie sense of time stretching out over hundreds of generations before me. We climb onto the mound and look down at the valley in which we live, which has sheltered so many lives before ours, which will hopefully shelter so many to come.

As the kids have their picnic, I read on my phone about the site underneath me. It is apparently built to align with the sun on solstice morning, the only day of the year when light will shine right to the back of the chamber. I make a mental note to return. The rounded, trapezoidal shape of the many long barrows that dot the Cotswolds is thought to be representative of a womb. Half buried in the earth, the ultimate womb which births all life on this planet, this planetary organ into which all life eventually rots again, to fuel the generations to come.

I am still trying to absorb and digest what I have recently learned about the generations that came before me. Sitting on this site marking the cycle of life and death, on this day when the veil between the living and dead is at its thinnest, I look out over this valley that is now our place. The horizon reaches out all around me, and a steady wind blows my thoughts clear as I remember my long search for spiritual guidance to help me find my way home.

In my adulthood it had increasingly struck me that all that I had been taught to believe to be true were ideas formed by the colonial dogma in whose miasma we are all steeped. That all I had been taught to contemptuously dismiss came from Indigenous belief systems that a supremacist Empire had tried to crush. I realised how closed my mind had been, how unwilling I had been to play with new ideas. I made the decision to open it.

I started near the medicine that I knew. I learned about Traditional Chinese Medicine, and had acupuncture, and felt awed by the twanging sensations my body felt when an experienced practitioner adjusted needles inserted along supposedly non-existent meridians. I had cranial osteopathy to adjust my pelvis when my baby would not engage towards the end of my first pregnancy, and laughed hysterically in the car for twenty minutes afterwards – in release or just straightforward amusement at what I had experienced, I could never be sure. I woke up the next day with my son's head fully engaged. I read Jung, outside the curriculum of my evidence-based medical psycho-therapy training, and was astonished to find references to astrology in my own field of work. I had always been fascinated by the stars, so I saw an astrologer for a reading that I hardly understood. From the stars to the earth: I began to foray into herbalism, spiralling back close to the medicine with which I had started, but in a much more earthy place.

Through all of this I had the sense that I was sharpening my intuition, honing my instincts – the same ones that now serve me so well in the garden. I threw off my attachment to old beliefs and opened my mind to imaginative possibilities. I allowed myself to wonder: what if?

I began to let myself be guided by gut feelings. The same ones that I relied on in my therapeutic work to inform me of the unspoken communication that lay in the transference, the same ones that I dismissed in my everyday life. One so clear as to be a command landed in me on a remote beach on Trinidad's north coast on my thirty-eighth birthday. I sat on the warm sand, my feet caressed by the swirling ocean, as childhood friends laughed and talked on the beach behind me. I watched my children play as I had done on this beloved wild coast in my own childhood. A life come full circle. I felt a tangled thread in my heart slip free, then pull closed in a knot of longing. One word filled my head with profound resonance as I gazed out at the constantly shifting ocean that lay between both of the islands that claimed me: *home*.

As that word and all that it carried resounded through me, it started us on the final leg of our journey. The urge that landed in my heart was an invitation to freely choose. To listen and let myself be guided by the sweetness of the land calling to me. To know that I was not stuck, that I only needed to take the first step and I would find the path I had not known that I was seeking laid out before me. It was a chance not to be forever trapped by happenstance or constraints, but an offer to gently drop my chains of fear to the earth, lay myself on its bosom, and be softly welcomed.

I stand with the children on this hallowed day at this sacred site that has existed for thousands of years. Feet strongly rooted onto the earth, I open my arms into the oncoming wind and let my mind soar into the valleys of possibility below that hold our home.

Salvia

THE NOVEMBER FOG has come. Every morning it swallows the valley. As we begin our days I peer through the kitchen windows into the eerie murk that surrounds us, finding the world altered, as nothing is visible beyond our own boundary hedge. It is as if during the night we have been plucked from the ground and now sit alone, on our little plot of home, floating among the clouds. The muffled mornings padded in cotton balls of mist are surreal.

I walk through the house in slippered feet and venture out onto the slippery deck to immerse myself in the strangeness. The air feels soft on my cheek, not the warm caress of the sun that I embrace, but gentle and pleasurable anyway. I have spent so long rejecting the presence of grey and wet, expecting the harsh rebuttal of unwelcome cold in return. Today I sink into the silky, enveloping smoothness of our valley. My skin is taut from the air of the house, which has begun to dry as the central heating senses the increasing cold. This barrier that contains, and protects, and connects me, softens in the fine droplets of mist suspended in the air. I breathe them in and feel my whole body blur into the landscape. Seventy per cent molecules of water, immersed in this moist air, a sea of atoms more tightly bound to one another than to those in which they swim. I dissolve, the ties that bind me dissipating with a hiss into the subatomic space that surrounds, silencing the noise of my thoughts. All the attachments that shackle me to myself melt away.

I am a world within worlds, and in this deliquescent moment I remember it: the 'I' that I am, made up of trillions of organisms – bacteria, viruses, yeasts, more – that colonise me inside and out. My 'I' cannot function without them, skin, gut, immune system dependent on the me-of-them to maintain my animal being. My cells powered by mitochondria, bacteria-like organelles which some scientists propose came from a symbiotic relationship between bacteria and multi-celled organisms that became merged at some long-distant time, integrated into our bodies. The DNA of these internal power stations, unlike the rest of the code that writes us into being, is passed solely along the maternal line, originating from our first mother bacterial.

I delve deeper, between the atoms that make up the molecules that create all the my-and-not-my cells. Atoms perpetually recycled since the dawn of this universe, the beginning of space and matter and time. The atoms that form me coming from the air I breathe, the food I eat; the longer I live here and eat the stuff of the garden, and breathe the air of the valley, the more this place forms the very stuff of myself. The longer I walk among the space and shed the dust of my self onto the ground around me, the more the place becomes the stuff of me. I welcome our mutual restructuring.

These atoms were previously conjoined with others,

then bonds broken and reshaped into me. The previous partners of my atoms – where now? But perhaps they are still influencing me, quantum entanglement showing that an atom somewhere else in the world once physically connected to one of mine will resonate with its previous partners' changes in state. Is it that which drives the inexplicable changes that sometimes hum through me? Is this how we sense things beyond sensing, collective waves of atomic energy pulsing through our forever interconnected selves? Is it this that brought me here, that creates the sense of something so familiar in a place so new?

I have reached the expanding outer limits of the realm of my comprehension. I coalesce again through my animal self, but it is still not individual, alone. Every other human being that was conceived and implanted in my body, even if only temporarily, remains within me. The developing placenta is porous, and some of the first foetal cells break away from the developing body and cross into the maternal one, where they remain for the rest of that mother body's life. No one knows for sure what those cells are doing in the maternal host body. Do they harm? Do they heal? My son and daughter, and other children unborn, live on in my liver, lungs, heart, brain. I am no one body. I am a chimera.

A gentle breeze stirs in the air around me, brushes against my cheek. I am wholly myself again, but reminded by this liquid morning of all the ways I am so much more, so much a part of this place. The sky lightens further, and a row of perfect webs on the balcony railings becomes visible, picked out by tiny droplets condensing from the moist air. The fog behind the trees warms to a soft pink, heralding the sun rising. I take in a deep breath, feel my lungs expanding. Time exists again. It is the start of a new day.

***

I have been thinking about spring. We are deep into autumn, but harvesting the fruits of this growing season compels a reflection on past months, and thinking ahead to the future. Time has felt very liquid this year, the days morphing into strange shapes, simultaneously excruciatingly long and bewilderingly fast. It has felt akin to the surreal early time of motherhood, where I first fully felt how time is not a fixed dimension, rather one that moves according to our perception. But where those hallucinatory, sleep-deprived, longest-shortest days were deeply personal, leaving me feeling isolated in a different time zone to the world carrying on at normal speed around me, this has been a mass effect. Despite our isolation from one another, there is a communal sense of how warped and confounding this time has been.

In the garden, time has taken on a shape different to the straight track that clocks and calendars and alerts on my phone mark it out to be, but it is comforting. Though I am new to gardening in this temperate zone, it all feels strangely familiar. Past, present and future exist simultaneously on each patch of soil. In a conversation over the fence with a neighbour who has lived in the village for many years, and watched our home and garden evolve before we ever set foot in it, I learn about a previous gardener who had the terrible habit of going out into the beds right after heavy rain, when the clay soil was saturated. I understand now the compacted, airless ground that I saw earlier this year, the impression of those feet marking the space still.

And yet the garden already bears my mark. There are seedlings, a green fuzz of germination on the still-warm autumn soil. Next year's plants the legacy of this year's work, all visible to me in this present moment. On the earth, time is a spiral. So much can happen in a growing season.

It feels familiar because this is how time works in the mind too. In therapeutic work, the same issues often spiral round to be encountered again and again. But when the work is progressing, often at a different depth each time. Things are the same, but they are not; the perspective has changed, the internal landscape has subtly shifted. This flux is healthy: it is when the mind becomes stuck in rigid ruts that problems arise. Disconnected from natural cycles.

I think about a book I had recently read: *Losing Eden: Why Our Minds Need The Wild*. In it, the author Lucy Jones beautifully presents reams of scientific evidence of why staying in close connection to the natural cycles around us is vital for our mental wellbeing. It made for stark reading, but also struck me as mad that we require studies and evidence to remind us of our true nature, when it should be instinctual knowledge of our animal species. The requirement to prove to ourselves how much we depend on this basic need is evidence of our collective insanity. And I wonder about what following this evidence to its conclusion might tell us – not that getting back to nature will make us more well, but that in having been disconnected in the first place we are all profoundly ailing. We have all gone mad.

My mind cannot fully hold the spiral of all that feels incomprehensible in the world. The issues of colonial conquest, or the racism that it bred to further and uphold itself, of the inequality it engendered, of the exponential extractivism it created that fuels the climate crisis – they swirl round my brain like a tornado. Each issue linked, each feeding the other in an endless, accelerating feedback loop. When my mind gets sucked into the circling updraft, I begin to lose myself in a hurricane of despair.

The only way out of that mad cycle is to root myself deeply into the rich earth. To rejoin with the interconnected hyphae of organisms that make up the ground that holds me. And in that landing, to feel the sigh of relief in every cell, the reigniting of a vibrant creative urge from a seed that had lain dormant under the deadened, compacted soil of my old life. The only way out of the spiral that binds us more and more tightly in its strangling interconnected threads is to touch the earth, to breathe into its grounding cycles.

And to feel the earth's uncanny relief at my return too. To witness its ecstatic explosion of beauty, feel it pregnant and laden with bounty in satisfied response to my clumsy ministrations. This moves me. I am no interloper here, no lethal virus, no foreign invasive. I am nature. I am Eve before the Fall. It was not the garden who cast her out. The garden loved Eve. Love is our birthright, but it has become perverse. It is uncanny how difficult it can feel to be truly loved.

And so in autumn I remember the past spring, and think

about future bulbs and seeds of life, and hope. The garden breeds abundance. Other more experienced gardeners in the village have been offering up their excess to me, and the kind generosity that gardening yields feels like love.

We are given daffodils and camassia from different meadows which have been thinned, and *Valerian officinalis* seedlings that have spread themselves about too abundantly. Many seeds, of poppies and hollyhocks, astrantias and heritage lettuce. They come from the gardens that surround mine, spilling out in immeasurable plenty across the constructed borders of fences and hedges. But they also come through the post, seeds sent by strangers I have connected with online. Each one when planted will be another loving root to ground us here.

I tuck the packets of seeds into next spring's seed box. I place these freely given bulbs in my garden's present soil, already changed from the past of our arrival. I plant them around the garden, dreaming of our future.

***

The season of colds and flu, enhanced by a resurgence of Covid, is upon us once again. I wander round the garden, many of the wild herbs still lush enough to yield something for my teapot, and make teas using plants beneficial to the immune and respiratory systems. As today's pot steeps, it occurs to me that, despite the summer's florid hay fever, I have not been seriously ill this year. In the last few years of living in Oxford, I had been suffering from increasing bouts of bronchitis and chest infections, ending up on antibiotics and steroids time and again. My constricted breathing had earned me a diagnosis of asthma, and an inhaler that I needed to reach for whenever illness hit.

This term I have been afflicted by the same colds that the children had on their return to school, feeling as if the viruses were more vicious for their hiatus from general circulation. But they had stayed bad colds and not troubled my lungs or affected my breathing. I had not needed to see a doctor since moving here. I take a deep breath, inhaling the steam of the tea and the sweet air of the valley. I am struck by all the ways that in this home I can breathe more freely, by all the levels on which this garden is healing me.

As I, perhaps, am healing it. I look out at the beds that were bare until my intervention. Now, even on the descent into winter, the space is abundant, covered in seed heads on which goldfinches perch, the soil blanketed with the decay of this year's growth. The return of life is a pattern that keeps repeating itself throughout the garden. The rhododendron in alkaline soil that did not flower at all in our first year, but after my touch this spring was covered in blooms. The dying *Magnolia stellata* that I found languishing, and replanted, which greeted this spring with a constellation of stars. The previously bare lilac that I mulched, this year almost groaning beneath its redolent weight. I cannot credit horticultural skill for these changes – this is my first English garden. I garden with little formal knowledge, but rich in instinct and love.

Sometimes when Oli and I talk about these changes, we toss around potential logical explanations. Perhaps there was weedkiller or some other toxin in the soil that has finally been washed out? Perhaps mulching really is miraculous? Perhaps. But one thing rings true again and again in my mind. When I take my limited body and offer some of my life to the garden, the garden reciprocates with life of its own. When I offer my love to the garden, the garden loves me back.

***

In the oddly mild autumn we have had so far, much is still in late flower. The most spectacular is the new bed that we created out of the muddy destruction of the lawn at the bottom of the garden. We will sow the rest of the space with the native wildflower seeds that we have sourced in the spring, and convert it into the lost meadow habitat that we keep reading is vital for insect life. But when a load of plants generously landed in our lap, we found them impossible to resist.

At the height of summer I had gone to her place following the lure of free plants. A message about an intriguing old nursery outgrown, with stock that had to go, that was sent to the newsletter of a local plant society that I had joined. It piqued my interest. I responded, thinking of the mudscape that we needed to rehabilitate now that the heating works were done. I received some dates in return, a map, a catalogue of daylilies. I looked up the place to which I would be journeying, learned that she was a Lady, an author and a botanist. I felt intimidated then, and wondered how she would greet my amateur incursion to her garden.

We met warily at first. Had polite exchanges about the

necessity of masks, gloves and the very unsocial distancing, then got down to the business of plants. Her garden was beautiful, wild – full of weeds and wildflowers which she told me were her specialty – among the carefully labelled overgrown clumps of hemerocallis and other plants. She told me of the horrors of publishing her book, the pleasures of giving talks at local plant society meetings, sadly suspended. I admired a colourful display, told her it reminded me of the colours of home. I told her why we were in search of plants, about our frozen winter of ground source heating installation and the sea of mud left in its wake. She was generous. Plant reparations.

My boot full, I retreated to the nearby shore, walked to the beach, stripped my hot, sweaty feet and dipped them with relief in the cool water as the winds blasted through me. Refreshed, I sat and checked my phone. There was an email from the Lady. What a pleasure it had been to meet me, what a shame that we could not have sat and had coffee, as she would have loved to have heard the tale of how I came from the Caribbean to my garden. I smiled, and then looked at the sea singing its siren song of life and death. I was at the edge of the Bristol Channel, and unbidden, like Colston's toppled statue slowly pulled by its feet from the murk at the harbour bottom, thoughts rose from the deep of the enslaved people at its depths, of my ancestors of all sorts who survived the sea's crossing. I sat on that rocky sister shore to the one of my childhood, as my feet dried in the stiff onshore breeze, and the crashing roar of the waves wiped my mind clean, and I was grateful to be alive.

Today I walk along the edge of the stream running

behind the new bed that we have created. I hear a familiar whir of wings behind me, then a trilling song. The garden's robin is back. It had followed me around in the spring, very interested in my mulching and digging, and what prized goods I might turn up for it. In summer I hardly saw it, busy with raising a flock of fledglings. But here it is again, now that it is getting on to winter. Right on cue.

Along with the daylilies, daisies and kniphofia that were given to us by the retired nursery woman, we have planted some treasures that we bought on our trip to Special Plants. The planting will need refining. At the time of creating this bed we just wanted to get the divided and uprooted plants into the ground as soon as possible to help them survive the summer. But transplanted into this unfamiliar soil they have thrived more than we imagined possible, and many are now in unseasonably late bloom.

Deep pink and maroon daylilies, to which I was drawn by my playful idea of creating a garden of mostly edible and herbal plants. I cannot bring myself to taste the flower buds yet; they are so beautiful. Maybe next year when the plants are bigger. A couple of tall spires of neon orange kniphofia, very tropical in feel, one of my husband's favourite flowers. Cerise lobelia, and a diaphanous verbena that I had never seen before, put on a vibrant show. But the most spectacular among them is the towering, deep black and blue *Salvia guaranitica*, the hummingbird sage. The blue flowers have an unearthly glow against the creeping brown of the autumn garden. I pluck one of its largest pungent leaves and, rubbing it between my fingers, lift the released aniseed scent up to my face.

The robin thrums nearer and associations hum round my mind. Hummingbirds of Iëre hovering in the garden; aniseed added to the herbs my grandmother boiled to make tea. The bitter drinks for my cooling, as I stand on the edge of the cool stream, roiling with autumn rains. Admiring the salvia while sage burned, its smoke waved around my grandmother's house as a clearing, an Indigenous healing ritual that Christianity had not converted away. This garden, that one, my children, my childhood, my mother and all my grandmothers and myself, the present and past entangled and rooted deep in this garden's soil. So much that was unexpected rises to the surface here. In touch with the earth, I spiral through layers of healing.

*****

The weather is unsettled. The rains wash in, followed by bursts of sunshine. Sometimes they both happen together. 'The Devil and his wife fighting,' the old saying I remember from my childhood about this contradictory weather. Something about the slanting autumn light, the frequent coincidence of sunshine and rain, and the shape of the valley, comes together to make this place magical. We discover that we live in the home of rainbows.

Nearly every day we see them, often doubled, soaring over the wood to end among the trees, or in the field, or on our neighbour's rooftop. One day, the rainbow seems to land in our own garden, on the veg beds at the back of the house where we have buried the garlic and shallots

— living treasure. I run outside in the light rain and stand at the spot where I saw the rainbow end, eyes closed, wet face upturned to the sun.

Sunshine and rain, together creating a luminescence, the whole far more beautiful than its parts. A whole which allows us to see truly the beautiful spectrum of colour that creates each beam of seemingly pure white light.

As without, so within. In this world rain-soaked with pain, that we have been taught we must not see, through the very act of our avoidance we close our eyes to the light. Every dark dissociation, every shut off and unexamined repression, numbs not only our pain but also our capacity to feel joy. Our ability to love, to transcend, to feel bliss and awe, to imagine the beautiful new worlds that will free us. Part-numbed, wholly weakened, in our desperate attempts to avoid the pain we fear that we cannot bear, we lock ourselves further and further away from our love and our joy and all which might save us. Each one too fractured to hold the pain of all, we forget that no two cracks align. Our broken uniqueness becomes our strength, when together the holding is whole.

The rainbow enchants me. This year my life has been slower and smaller than ever before, but still I feel more full, more rich than ever. I am more connected to the pain of the world than I have ever been, and yet, in the wholeness of my connection, I am, for the first time that I can remember, content.

Following the rains, mushrooms appear everywhere. They are the decomposing garden's strange autumn flowering. They pop up in the middle of the grassy paths between the new veg beds, grow out of the side of sleepers, form rings on the bottom weedy lawn. Some of them I think I recognise – common inkcaps and waxcaps, jelly ears and possibly boletes. A brief explosion of slime mould. I do not touch any of them, wary of my poor skills at their identification, but crouch and gaze at them with wonder, the way they magically appear overnight. I read about fungi in trying to identify them. My mind is blown by the complexity of their function and structure, by how much our very survival relies on their lives carried mostly invisibly underground. I read about the fungus that invades ant brains and takes over control of their behaviour to disseminate its spores. I have never seen so many different mushrooms as I have here. I squint warily at the clumps blossoming in the damp garden and wonder how much I might be under their control.

The children find some bright green jelly-like balls glowing on otherwise poor, bare earth beneath a shabby-looking section of the beech hedge. Here the garden bed is extremely narrow, and there is little soil between the stone wall and road in which the hedge can grow. We look the gleaming balls of jelly up – they are blue-green algae commonly called witches' butter or star jelly, names as magical as they look. I am amazed to read that they can survive anywhere from the artic to near active volcanoes. This patch of scruffy land seems a tame choice for them, considering. I wonder if they are the same type that was used in the unpalatable green powders I mixed into drinks in the days of trying to heal my infertile body. They are nitrogen-fixing. The soil in the section of the garden where we have found them is particularly poor. I wonder if they will apply their healing powers here. The children harvest them and mash them with water from the stream, gleefully concocting a spell with their witches' brew. They tell me to make a wish over it. I wish that they never lose this sense of the earth's healing magic.

With the rains come more morning mists. One murky dawn, in the half light that is neither day nor night, I stand once again immersed in the mist on the deck. The call of a bird reaches me through the fog. I try to place it, but in this watery air it is hard to listen. Learning the lessons this land wants to teach me means listening, and sometimes I do not want to. Like the wild child that lives within me still I rebel and rage, or turn my back and, preoccupied with some bright, shiny distraction or other, ignore her gentle insistences. My reasons may seem many, but at heart it is always the same: I am vulnerable and afraid. How can I bear to learn these shameful lessons she would show me of myself? How might I find the courage to make the change she urges? It is beyond my human frailty; it is impossible.

And then I am filled with wonder as I remember. That, mired as I may feel in the mess of human contradiction, in truth I spin round and round faster than I can imagine. I am hurtling on this earth through space and time, divine voyager of galaxies. I am a galaxy, atoms untold vibrating in harmonious rhythm the miraculous making of me. All this and I am human flesh built of the molecules constructed by the animals and the plants and the fungi of the soil beneath me. When I lie on the earth, the gaping wound in my soul is the opening through which I root into this earthbody that holds me. I am the land. How can I not listen when her voice is my own, when the all-that-I-am is possible.

***

Leaves gather in corners of the stone steps, and I am amazed by how quickly they turn into compost. Every time I sweep, in an attempt to keep the paths clear and the patio from turning slippery, I disturb whole colonies of woodlice and worms making new earth from the garden's death. The garden is rotting. The energy that built up this summer's garden begins to dehisce, will compost into the soil, and then fuel the cycle of next year's growth.

I see a lesson in this for me. Constrained to the garden so much this year, I have looked on increasingly despondent at the world around me. Reading and re-reading the works of scholars and activists who have come before, feeling how little we have taken to heart the lessons of history, my heart sinks. I have a growing sense that, in directing my energy and attention towards beseeching old colonial institutions to try to do something new, I am merely participating in the old systems, perversely propping up and breathing life into that which I wish to die. I want to break free of these old cycles. The autumn garden seems to show me how.

I imagine myself dehiscing from these systems. Letting the old growth of colonialism fall from the trunk of my body and rot, systems starved of the fuel of my attachment and attention. Pulling my energy back inward, down my phloem and into the roots growing right beneath my feet.

Outward further into the hyphae of the fungal bodies that will convert the dead old growth into nutrient-rich compost, that will generate the psychic energy that underpins all creative and regenerative instincts and urges. My libido – the alchemy of directing my vitality and love into creating something new, right here where I stand. Growing something strong and enduring at last. The seed case of my defences cracked, the soil improved by my composting, made receptive. Expansive growth can follow, now that I have put down strong roots.

I have grown into my name: Ayana. They named me 'beautiful flower' in the language of imagined ancestors. I used to think this a mere pretty frippery, but as I stand in this late autumn garden, for a moment I finally understand the power of my naming. It reminds me that I am no meaningless charming posy, but gifted with the alchemy of life. The ability to transmute the shit of suffering that went before into a new creation, shining beautiful and true. That I am here in glorious fecundity to be pollinated by ideas anew. Here to set seeds of hope on this restored soil of the future. We all grow in relationship to one another, and so many relationships need repair. Together with this place something has been mended within – inside my self and my garden. It is time to turn outward.

I say yes to becoming a school governor, take up a volunteer shift at the community-owned village shop, put out a notice in the email newsletter about a meeting for those of us who would like to work towards improving our village's ecological future. I begin to write a column for the school newsletter on coping through the ongoing pandemic. I sit in front of the doors to the deck, suspended between earth and air, gazing over the woods. This feeling is freedom.

As a child I wanted to be writer. I was told for the safety and security of my own future that this was not viable, encouraged to pursue my interest in human stories through the path of medicine instead. I look at the garden, and I begin to write.

***

It is a sunny, perfect day, so we build a roaring blaze outside. The mild weather has suddenly passed and my fingers sting with cold. The children light sparklers and toast marshmallows and dance around the fire. We are on the descent to winter once again and the darkness looms even larger this year than last. I think that I really must order some more mulch.

The next morning, I wake up to a frozen world. The first frost has arrived; winter's shining kiss. Every leaf and blade and web caught in sparkling animation. With childlike delight, I pull on a thick coat, gloves and wellies and spill out into the garden with the children. Leaves that were sodden and ugly in their decay yesterday shimmer in the glitteringly cold morning. All the late flowers hang suspended in ice. How beautiful it is to watch a summer die. How necessary this death to next season's life. We look at the intricate patterns traced out on the crystalline leaves, laughingly follow the clouds of our breath. The truth feels sharp and clear and simple and bright on days like this. I love this place. I love who it allows me to be.

Epilogue

Viburnum II . . .

OUR ANNIVERSARY OF living in this home spirals round. We spend it in the garden. We bought bulbs in the December sales, and spend the day moving around the beds, burying them as quickly as we can in the rapidly cooling earth. Still warm enough to heat our home – our ground source heating system has kicked in and is working a treat. Sometimes this system of pipes from the earth warming our home seems unfathomable, like a mystery from the deep. I think of the 130-metre-deep roots that we have put down here, and hope they will anchor us in the climate storms to come.

With my tray of cut-price bulbs, I walk round to the front of the house in search of a bare patch of earth to liven in the spring. The viburnum is in flower again. I smile at its tiny pink bunches, pick one and put it as a scented posy in the top pocket of my dungarees. Its warmly sweet scent is redolent of home. Of all my homes.

A message pings into my phone. I sit on the front doorstep and read it. It is from my cousin Charmaine, who has remained in contact about the research she has been continuing into our family stories. This one seems to be a compilation of other information she has already shared. I scan it quickly, expecting no new revelations, and start mentally composing a brief reply, when something catches my attention. She has given my grandmother a middle name, one that I have never seen before. I check the time – she will be awake now – and quickly call my mother to confirm. Yes, Mama had a middle name, didn't you know? It was Daphne.

My breath stops. I look at the *Daphne odora Aureomarginata* opposite me. It has grown hugely since I planted it, now many times larger in size, a low mound of glossy green edged with white spreading towards the trunk of the wisteria. I notice the first signs of flower heads beginning to form at the top of each divided stem. I remember how I spotted it in the garden centre shortly after our arrival here, how it felt destined for this spot next to our front gate. A wave of goosebumps covers my flesh. I thought I had found her memory in the roses, but I have planted her here. My grandmother blooms in my garden.

***

The winter solstice comes. At dawn we walk to the Long Barrow to welcome the returning sun. I hope that I will witness its alignment with the sunrise, and that we will be blessed with some magical epiphany as light strikes the very back of the tomb. But it is overcast, and it seems the tomb is imperfectly aligned, and apart from one brief but awe-inspiring moment of the sky being lit fuchsia pink by the rising sun, there are no moments of cloud-parting significance.

But we have come with a group of other adults and

children from the village who have grown into good friends over the past two years. The children, coaxed from warm, dark beds up the cold hill by the promise of hot chocolate and cinnamon buns, run and tumble across the mound, their laughter mingling with the drums emanating from the pagan rituals happening within the tomb below. Our group sits to one side, watching the sky lighten, sharing breakfast, stories, songs. It had seemed a slightly unappealing idea when I first proposed it, to usher a group of toddlers and young children up the hillside in the winter dark, but washed in the light of solstice morning, warmed by laughter, we are all happy we have come.

I recognise someone from a group of adults standing nearby; another villager. We wish each other happy solstice, and she laughingly tells me that the group she is with have been coming here every solstice dawn since their children were the age of ours. I stare at the adults smiling on the hill, then look back at our young group. A wave of tenderness and gratitude that we have found a community to root into, that I have been able to let myself entangle with others in this place, washes over me. As we pick our way down the valley back to our houses, we resolve to make this an annual ritual of our own. New traditions blooming on home soil.

***

After Christmas, Oli and I sit in the conservatory one evening after dinner, enjoying the glow of candlelight while the children run up to their room to play for a few minutes before their bath. The flickering light reflects off the dark windows above and around the multiple selves that look back at us. For a moment I can feel all the 'we' who have lived in this garden.

I am in a strange mood; everything feels eerie. The weather has been grim, a wet, dank Betwixtmas, but today I managed to get out into the garden in a brief break in the rain. I pottered, weeded a bit, checked up on the plants in a new area that we created this summer. The tar-impregnated sleepers that had been holding up the ragged bed of shorn bamboo had finally collapsed. In the creation of our first bit of hand landscaping, the poisoned tropical wood had been buried and used as retaining structures under new local-stone walls. I had thanked the sleepers as they were being stacked under the new curves. The garden's past would provide a strong foundation for the future we were creating.

While in the garden this morning, I saw signs of life – bulbs emerging, life gradually awakening in the imperceptibly lengthening days after solstice. The life that had felt so absent from the garden in our first year making a welcome appearance. The winter garden is still so lush – standing seed heads and thick clumps of overwintering grasses, the evergreen rosettes of perennials biding their time to burst forth into new growth in the spring. I could feel it there, pulsing in the cool earth beneath my fingers. It is winter, and it is not death. I feel the rise and fall of its slow breath as I caress the blackly mulched soil of this new planting. The garden sleeps this winter. We have loved it back to life.

Over dinner, we were reminiscing about what led us to choose this place, and I gushed to Oli about how much

I had loved the garden. It has taken me three years of living here, of loving our old, wonky house, of feeling deeply comforted by the way that every surface surrounding us when indoors is made of stone or wood, to realise that while I was looking for a home to root into, I was never looking for a house, but a garden.

I remembered aloud what I had seen on our first viewing of the garden, how lush it was, and Oli gave me a strange look, pointing out that the bones had been beautiful, but it had a sense of being quite bare the autumn that we looked round. I laughed in disbelief, remembering how luxuriantly lovely it all seemed. He pulled up the old estate agent particulars and showed me the pictures of the garden that had lured us to see it, old photos of the land captured on Google Maps. I could not believe what I was seeing. The desolate landscape that had distressed me so much that first winter was the reality of the garden we had viewed and bought. How could I reconcile that with my memories of the place?

Memory is not in the past but in the present. In our unconscious mind there is no past or future tense, everything happens in the perpetual now in which our lives are lived. Every time we remember an event, our brain reactivates the occurrences and treats them as if we are once again living through those moments in the present. But not *as if*: we are once again living through them in the present. Every remembering rewrites the story afresh.

Sitting in the conservatory in the flickering dark, surrounded by layered images of myself, I look at the photos on my phone of the garden this summer, three growing seasons on. The beds are lush and wild. I had taken these pictures at the high point of summer, when the garden was thick with clouds of insects, its growth feeling as if it had tumbled completely out of control and would overwhelm me, a familiar feeling that I recognise now as coming just before it begins to sink into the slow wane to autumn.

Looking at the phone, I realise that this summer's lush abundance, far outstripping the first year's growth that had so amazed me, is the garden in my memory of our first viewing. Somehow, when we walked round the space that first day, I had not seen how the place before me actually looked, but what it might be, what the garden might become with us here. Present sight overlain with the memory of the garden's future. I am lush and wild, no longer the dry, barren creature of our arrival. The garden had seen what I could become while living here. It had showed me what we could grow into together.

I take a long, deep breath. Inhale an epiphany. Sitting among the houseplants in the conservatory, I can see how my early root system was shallow, easily disturbed, on unsettled ground. I can feel how I had been like a plant cutting in an increasingly murky glass of water once I arrived in the UK – free-floating and untethered. But in this garden, uprooting the noxious weeds of colonialism that choked my life, I cleared fertile ground receptive to me, and me receptive to it, and this mutual love allowed me to grow deep, strong roots at last.

The future remains dark and uncertain, and while the future is always uncertain, it seems that in the events of recent years the veil of the collective illusion of certainty and stability has been permanently torn. Who knows where we will find ourselves next year? Yet along with the plants that have showed themselves to me, and me to myself in my garden, I know that, once rooted, I will thrive.

I go out into the dark garden. I touch the earth, the earth touches me. The living hyphae on my skin join the living hyphae on the soil, my atoms melding into the matter that upholds me. I am a creature of this earth, and for a moment I feel a charge of love surge through me, down through my fingers, and back into my home.

Acknowledgements

It TAKES A village and I am so grateful to all of you who are in mine. This book has grown through the loving care of so many from well before its inception.

Rebecca, I can never thank you enough for reading my essay about my garden and my grandmother, and in your incredible way seeing everything that was to come from that small seed, and making so much magic happen, including introducing me to my wonderful agent. Julia, you are the best agent I can imagine. Thank you for seeing in me what I absolutely could not, and for shifting my entire perspective of myself and what is possible. Sarah, you also came into my life through Rebecca's magic. Thank you for your patience and generosity and gentle wisdom, and all the enjoyable hours of writing therapy.

To the judges of the Nan Shepherd Prize, thank you for your faith in my writing. And to Alice especially, thank you for the final nudge of encouragement that I needed to apply. Caro and Megan, I don't think you'll ever believe how much your lovely email changed my life. The incredible scenes in the village hall when I opened it!

Aa'Ishah, thank you for being the best editor a debut memoirist and completely green writer could hope for. Your thoughtful and sensitive handling of my work (and me) has been such a gift. Helena, thank you for taking up the baton so ably when it was passed on to you. Lorraine, thank you for your careful reading and buffing the rough edges off my words so that they shine more brightly. Gill, thank you for creating such a beautiful vision to clothe my first book! And to the rest of the Canongate team, every single one of you that I have interacted with, you are a garden of pure delight. I'm so lucky to have landed on such rich, supportive soil.

To my virtual commune, who have been there lifting my words up since the days it was cool to blog anonymously on the internet, may a million memes of loving thankfulness always flood the group chat. Thank you for being my first readers, and always being my people. To my more recent readers online, thank you for your kind encouragement, which helped me see that my seedling efforts at writing could bloom if only I dared to nurture them. And to the rest of my posse, who know my words mostly as they are spoken, there since school and uni days, thank you for your unsurprised but still excited acceptance and unwavering support of whatever move I pull next.

To my real-life village, that holds me, my family and my garden, I wrote a whole book to try to express my love and gratitude for you all, and yet words are not enough. You welcomed me during the most surreal of times; you feed and care for my children, make me laugh raucously and have reinstated my faith in humanity. You make me feel so profoundly at home. You are family now! I cherish you all so deeply and I am so glad my life is rooted alongside yours. May we all flourish and grow old together.

To my actual family, thank you for endlessly, patiently, and lovingly putting up with me. Thank you for making me who I am. Annabel, thank you for helping to mother my children despite everything, so I could find time to write. Adrian, I'm so glad you were able to hold a copy of the book. You will be forever missed.

Mom and Dad, you are simply the greatest. Thank you for always supporting me in everything, and for giving me what I know is more than I will ever be able to fully appreciate or understand. I love you. I love that you're so very proud of me! And I hope that wherever her spirit is now, Mama is too.

To my son and daughter, thank you for bringing me on the most amazing, life-changing adventure that is motherhood. Thank you for making me brave. I wrote this for you.

Oli, your heart is home. You make everything possible. Thank you, thank you, thank you.